GREAT CIGAR STUFF

STUFF

FOR COLLECTORS

Schiffer Publishing Ltd

4880 Lower Valley Rd. Atglen, PA 19310 USA

JERRY TERRANOVA &
DOUGLAS CONGDON~MARTIN

DEDICATION

Jerry dedicates the book to his family, and Doug dedicates it to Tom Gray of Winston-Salem, North Carolina, for his support and friendship.

ACKNOWLEDGEMENTS

We would like to give special thanks to Larry Shapiro, a dealer from Glastonbury, Connecticut, who lent his expertise to the task of assigning current market values to the items shown in this book.

Designed by Bonnie M. Hensley

ISBN: 0-7643-0368-6
Printed in Hong Kong
1 2 3 4

Published by Schiffer Publishing Ltd.
4880 Lower Valley Road
Atglen, PA 19310
Phone: (610) 593-1777; Fax: (610) 593-2002
E-mail: Schifferbk@aol.com
Please write for a free catalog.
This book may be purchased from the publisher.
Please include $3.95 for shipping.

In Europe, Schiffer books are distributed by
Bushwood Books
84 Bushwood Road
Kew Gardens
Surrey TW9 3BQ England
Phone: 44 (0)181 948-8119; Fax: 44 (0)181 948-3232
E-mail: Bushwd@aol.com

Please try your bookstore first.
We are interested in hearing from authors
with book ideas on related subjects.

CONTENTS

INTRODUCTION

From all the hoopla it may seem the cigar was newly discovered in the early 1990s. On every street corner, it seems, there is a new cigar store, its walls lined with humidity-controlled cabinets designed to protect the precious and costly commodities within. Restaurants and clubs advertise "Smokers," special nights when men and women gather to share a good cigar or two. Magazine racks carry not one, but three or four glitzy magazines devoted to the smoking of cigars and the accompanying elegant lifestyle. And, yes, there are more than a few books devoted to this new-found indulgence.

Actually, what seems "new-found" is a rediscovery of an old pleasure. The history of the cigar goes back at least as far as the Nina, the Pinta, and the Santa Maria, when natives were spotted smoking firebrands by Columbus and his fellow travelers. The intoxicating pleasure of tobacco was not lost on the explorers, and before long Europe was hooked on the aroma of burning leaf.

In the United States joys of cigar smoking did not reach popular levels until the Civil War. Preferring a good chew, a pipe, or, especially among the ladies, a touch of snuff, it took the cigar chomping heroes Grant and Sherman to awaken the populace to the cigar's particular charms. Once started, the cigar's popularity continued to rise well into the 1920s, when the cigarette began to overtake it. By the end of World War II, the cigarette was the unchallenged smoke of choice.

Cigarettes have maintained their position to the present day, but have come up against a growing mountain of scientific and political opposition. It started in 1952 when *Reader's Digest* published an article from *the Christian Science Monitor* entitled "Cancer by the Carton," which chronicled the health risks posed by cigarette smoking. In 1966 the U.S. Surgeon General issued an order requiring that cigarette packs contain the warning "Cigarette smoking may be hazardous to your health." The warnings grew stronger and gradually were extended to all forms of advertising. In the early 1990s the tide of anti-cigarette forces rose to new heights. Local governments across the United States banned smoking in public places. Stores and restaurants sprouted new signs declaring themselves "Smoke Free." Courts were jammed with litigations against cigarette companies brought by smokers, their survivors, and even non-smokers who believe they have suffered the ill-effects of secondhand smoke. In the spring of 1997 the Federal Trade Commission announced it would challenge the use of the Joe Camel character by the R.J. Reynolds Tobacco, Co., reversing an earlier decision.

Ironically, the quagmire that has engulfed cigarettes may be contributing to the renewed popularity of cigars. Since they are normally not inhaled, cigars have long had the reputation of being less harmful than their cigarette cousins. Despite warnings that cigars, too, carry some risks, for the health-conscious they offer a way of enjoying the pleasures of tobacco, while minimizing the danger. Once tried, with the rich flavor and aroma exciting the senses, a fine cigar is a luxury comparable to a fine vintage wine or an aged Scotch whiskey. A search of the Internet as this is written reveals nearly 100,000 sites that deal with cigars, a number that will no doubt grow exponentially.

It is little surprise, then, that cigars have given rise to such a material culture. As we showed in our earlier book, *Antique Cigar Cutters and Lighters* (Schiffer Publishing, 1996), the artistry, quality, and shear volume of cigar related paraphernalia says much about the value placed on this little pleasure. Indeed, the number of antiques surrounding the cigar is great enough that this present book is equally rich, covering a much broader range of items.

A Word About Prices

The prices in this book represent an educated estimate of what you could expect to pay for the item if found in an antique store in fine to excellent condition. Because the author's experience is centered principally on the east coast and prices vary according to geography, the buyer will likely experience some variation. While we have attempted to be accurate, neither the authors or the publishers can be responsible for any gain or loss experienced by using the prices listed. The buyer's most prudent strategy is to work with a reputable dealer and to shop around for the item he or she seeks.

We hope you enjoy a look into the past, and some of the things our love of the "firebrand" cigar has produced.

PART ONE:
CIGARS AT HOME

1. HUMIDORS, BOXES, CUPS, AND MORE

Nothing is more important to the pleasure of smoking than a good humidor. A good cigar is an investment in pleasure, and protecting that investment requires a friendly, stable environment. For the cigar that means a humidity of around 70 percent and a temperature of about 70 degrees. If the humidity drops to 68 percent or below, the cigar slowly dries out. If it reaches 80 percent, it is likely to develop an unsightly and unsmokable mold. While cool temperatures will not cause much problem, warm temperatures encourage the growth of tobacco worms. If conditions are right cigars can be kept for years, even decades, with little or no loss of quality. Indeed, some smokers prefer an aged cigar.

Besides being essential to good cigar smoking, the humidor can be a thing of beauty and value. Even putting aside the aberration of the half-million dollar humidor from the Jackie Kennedy auction, humidors can be quite costly when new, and some of the older humidors shown here have increased with value over the years. They include cabinets trimmed in ornate carving, and beautiful mahogany and brass chests with liners of milkglass or metal and elaborate seals to keep in the moisture. They also come in the form of canisters, some of which are beautifully designed with glass and metal trim.

While the humidor is vital for long term care, to work properly it should be opened as infrequently as possible. This means that cigars need other, more accessible containers. These include cigar boxes made as decorative accessories, as well as cigar cups, designed in many fanciful forms. As objects of beauty some of these are unsurpassed in the cigar world.

Finally, for those who like to carry a couple of cigars with them, some handsome and practical cigar cases have been designed. From enamel painted cases to molded aluminum, they were available in a style to suit everyone's taste.

Left and above: Carved wood cigar cabinet. A drawer in the base holds paraphernalia while the cabinet contains seven sliding trays holding eight cigars each. Manufactured by Coretz & Company, Manchester, England. 16.5" x 12.5" x 7.75". $2000-2250.

Left and above: Burl veneer and inlaid cigar storage cabinet. Holds sixty cigars on sliding trays. Circa 1880. 13.75" x 11.75" x 7.5". $1500-2000.

Oak and brass humidor chest. Beautifully engineered, the lid opens when the drawer is pulled. 7" x 13" x 8". $1500-2000.

Large Benson & Hedges humidor. Mahogany with brass inlays and hardware. The lining is metal with a copper seal. The humidor measures 13" x 25" x 14". It was meant to sit on a table with a large drawer for paraphernalia. The table shown is not original. $2500-3500.

Left and below: Mahogany humidor by Benson & Hedges. Mahogany with brass hardware and interesting brass inlays at the corners, it measures 10" x 21" x 12". $2500-3500.

Above and right: Benson & Hedges mahogany humidor, circa 1920. The plated copper liner is double walled to hold water. The flange of the top fits between the felt seals to keep the air out. Brass campaign style hardware is on the ends. 7" x 9.5" x 14". $1500-1550.

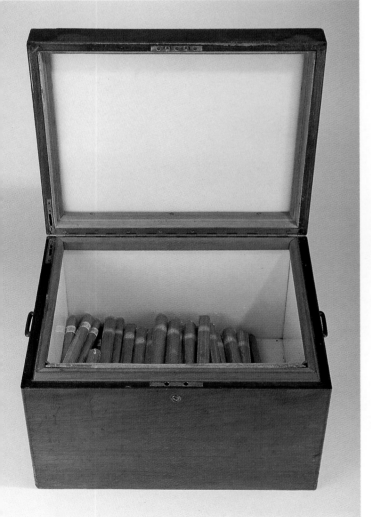

Left and above: Large mahogany and milkglass humidor with brass hardware, including campaign-style handles at each end. 10.5" x 16" x 13". $600-700.

Left and above: Milkglass and mahogany humidor, circa 1940. 5" x 11.5" x 7.75". $400-450.

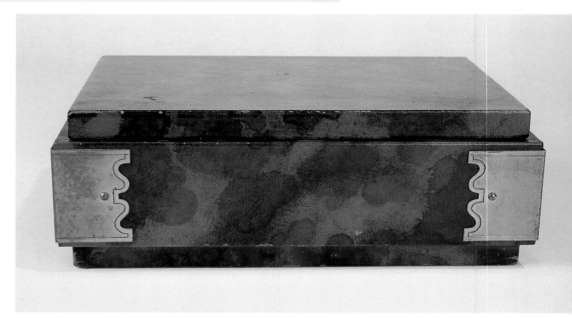

Left and above: Milkglass and cedar humidor with leatherette cover and brass fittings. 4" x 11" x 8". $200-225.

Left and above: Milkglass and mahogany humidor, circa 1940. Excellently executed with solid brass hardware and a deep, dark finish. 4.75" x 11.5" x 7.5". $400-450.

Below: Deeply detailed wood humidor with milkglass lining. 7.75" x 12" x 8.25". $500-600.

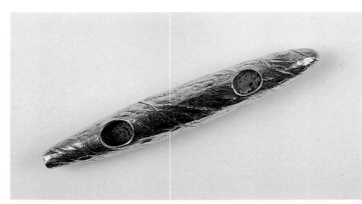

Left: Cigar shaped humidifier for a humidor. The sponge within holds water and gradually released moisture. 4.5". $75-100.

Below: Mahogany humidor with metal liner. 5" x 11" x 8". $400-500.

Below: Walnut and milkglass humidor with an interesting clasp. 4.25" x 11.5" x 8.75". $400-500.

Mahogany and milkglass humidor. 4.5" x 11.75" x 8.25". $400-500.

Small mahogany humidor with milkglass liner and a music box in its base. 3.25" x 8.25" x 4.5". $200-250.

Above and right: Dog house humidor. The carved wooden dog guards the cigars inside. The house is of wood with brass fittings. Missing are the two match holders that stood at the front corners. 6" x 6.5" x 11.5". $350-400.

Above and right: Bull dog humidor. The metal dog sits in a wood humidor house. The striker is in front of it, and two match holders are on the sides. 5.5" x 6" x 8.5". $250-300.

Silverplated cigar humidor by Queen City Silver, Cincinnati. It has two hinged humidor chambers with a chamber between for placing a moistened sponge. Holes between the chambers allow for the circulation of moist air. 4.25" x 8.25" x 5". $400-500.

Silver plated cigar humidor by Meriden Silver Company. The cigars go in the larger hinged compartment. Beneath the owl is a compartment for a moistened sponge. 6" x 6.5" x 5.25". $400-500.

Beautiful hinged humidor in silverplate. 7.25" x 7.5" x 6.5". $500-600.

Left and above: Ceramic humidor. 3.25" x 8" x 4.5". $300-350.

Brass Duk-It humidor by McDonald Products Corp., Buffalo. 6" x 4" d. $125-150.

German ceramic humidor. The beer stein reads "Gesundheit!" (To Your Health!). Marked Czechoslovakia and Made in Bohemia on the bottom, along with the initials JMD. 9" x 4" d. $900-1000.

Chrome "Rumidor Humidor" by the Rumidor Corporation, New York City. The knob at the top turns to "on" or "off," opening or closing the moisture vents inside. 8" x 6.25" diameter. $175-225.

Glass and silverplate humidor, possibly for Bull Dog Cigars, though it is not marked as such. 7.5" x 4". $275-300.

Bronze humidor with applied brass work on the lid and on the six panels. 8" x 6.25" d. $300-400.

Brass and metal humidor with leaded stained glass on the six panels. 8" x 6.25" d. $300-400.

Brass humidor with copper trim. The markings are Russian. 10" x 5" d. $275-300.

Home made copper humidor in the form of an artillery shell. 7.5" x 4.25" d. $100-125.

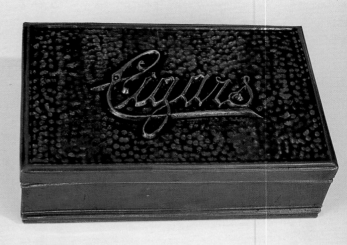

Copper plated cigar box. 2.5" x 8.5" x 5.5". $125-150.

Barrel-shaped humidor of wood with ceramic lining and brass fittings. 7.25" (excluding handle) x 5.75" d. $100-125.

Heavily embossed brass cigar box with sledding scene. Cedar lined. 2.75" x 9" x 5.5". $250-350.

Above and right: Cedar cigar box with brass hardware and finger joints. It contained 100 cigars. Factory No. 613. 3.75" x 14.25" x 6.5". $50-75.

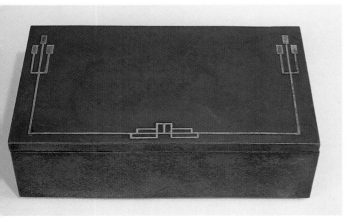

Art Deco box in cast bronze with sterling silver trim. The box is lined with cedar. 3" x 10" x 6". $200-250.

Cast iron cigar box. The figure in this variety has a match box in hand while in another variety he is holding a large cigar. 9" x 6" x 6.5". $750-850.

Ceramic version of the cast cigar box. 9" x 6" x 6.5". $650-750.

Older version of the figural cigar box. Cast iron, Larosa Cigars on the top. 9" x 6" x 6". $750-850.

A bear pulls a sled of cigars in this wonderful ceramic cigar box. Marked Bock Havana on the side. 5.5" x 13" x 3.5". $700-800.

Cast iron cigar box with elf and pipe figures on top. Nice polychrome finish. 6" x 6" x 4.5". $350-400.

Father Christmas cigar canister. Cast bronze with gilting. 7.75" x 7" x 5.5". $500-600.

Hand carved wood cigar or cigarette case in the form of a soldier's laughing head. Signed E. Seilberg, Ilykoping, 1942. 6.5" x 4.25" x 6". $300-350.

Cast metal desk piece with brass cups for cigars, ashtray, and matches. 5.75" x 6.25" x 5".

Cast bronze cigar box. The top of the chimney, which has a match cup, lifts off for access to the cigars. Signed G. Gardet. 8" x 4.5" x 3.75". $900-1000.

Left and below: Bronze Indian bust. Intricately sculpted, it has spaces to store tobacco, cigars, cigarettes, and matches. $7000-15000.

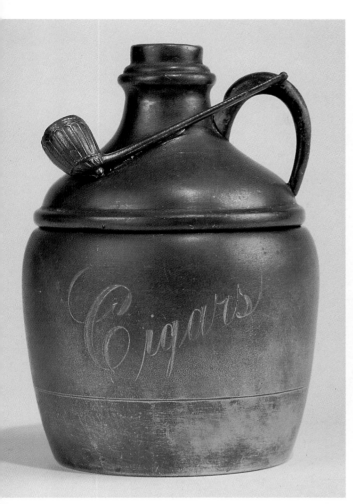

Left: This silverplate cigar canister has an identity crisis with a pipe on the lid and the word cigar engraved in the bowl. Imperial Silver Co. 6.5" x 4.5" d. $175-250.

Elegant silverplated cigar canister with attached tray. The lid is hinged. 7" x 8" x 7". $400-500.

Left and above: A bundle of cigars and a delightfully crafted lid mark this silverplated cigar canister. Made by James W. Tufts, Boston. 5.5" x 4.5". $400-500.

Though not cigar related this piece is of interest. Silverplated and nicely chased, it was designed to fit over a can of Blue Boar Tobacco. Undoubtedly a few cigar lovers had other uses for it, so we include it here. 6" x 5.75" d. $100-125.

Glass humidor, circa 1950. 7" x 5.25". $150-175.

Molded glass cigar jar. $75-125.

The champagne bottle was a popular shape for cigar containers. This one is nicely engraved with the word Cigars on the top and flowers on the bottom. Silverplate by Bristol Plate Co. 10.25" x 2.75" d. $300-350.

Champagne bottle cigar container. Silverplated, made by Pairpoint, New Bedford, Massachusetts. 10.25" x 2.75". $350-400.

Engraved champagne bottle cigar container. Hamilton Co., 10.25" x 2.75". $300-350.

Figural cast iron cigar and matchholder. Unmarked, it has a nice polychrome finish. 6.5" x 7.5" x 3.5". $375-450.

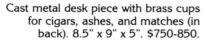

Cast metal desk piece with brass cups for cigars, ashes, and matches (in back). 8.5" x 9" x 5". $750-850.

Cast metal figural cigar holder. There is a cup for matches and at one time an ashtray was held in the figure's lap. Beautiful polychrome finish. 7.5" x 9.5" x 6.5". $450-600.

This wonderful lamp of wood and brass is a cigar and matchholder. The cigars were held in the cup to the right, and the matches were held beneath the dog. 21" x 9" x 4". $650-750.

Left: Delightful figural cigar and matchholder. Ceramic, 7.75" x 7.5" x 7". $275-350.

Left: Ceramic souvenir cigar cup. 3.75" x 2.25" diameter. $75-100.

Above: Two-faced bull dog cigar holder. Molded copper, glass eyes, and a brass cup. 3.25" x 5.25" x 4.25". $300-350.

This nicely detailed bronze porker is a cigar and match holder. 4.25" x 8" x 3". $225-250.

Plated cigar holder in the form of a top hat. Nicely chased. 2.5" x 4.5". $175-225.

Above: Bronze figural cigar holder. $250-300.

Right: Wood cigar stand and match holder carved in the Bavarian style. The stand rotates on the base. 10" x 6" d. $500-700.

Wood cigar stand and match holder with brass fittings. 6.5" x 6.5" d. $150-175.

Above and right: Lacquered wood and brass cigar cabinet. A twist of the knob opens the six doors, revealing three cigars in the brass rack behind each one. 9.75" x 7". $300-375.

Left and below: Handmade rosewood and ivory holder for six cigars. Each log will hold a cigar, accessible by lifting up the logs above. 3.75" x 6" x 3.5". $1000-1250.

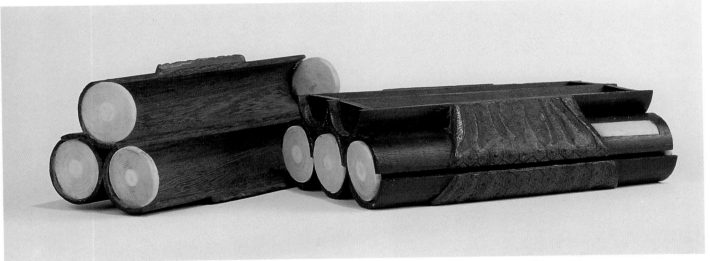

Brass cigar stand and matchholder, with cast base and molded top. 5.5" x 8.75" x 5.25". $175-225.

Apollo desk set in bronze and green glass. The set includes a matchbox holder, ashtray, kerosene lighter, cigar cup, and cigarette cup. $1250-1350.

Imitating some European enamels, this expandable case has a hand painted saintly lady on one side and a little more worldly picture on the other. Board with accordion expansion sides. 5.5" x 2.75". $350-400.

Another expanding cigar case, though this one is hand painted on only one side. 5.5" x 2.75". $200-250.

Hinged tortoise shell case with applied metal decor. 5.5" x 3.25". $450-500.

Expanding cigar case, hand painted on one side only. 5.5" x 2.75". $300-350.

Carved tortoise shell sliding cigar case. 5" x 2.5". $300-350.

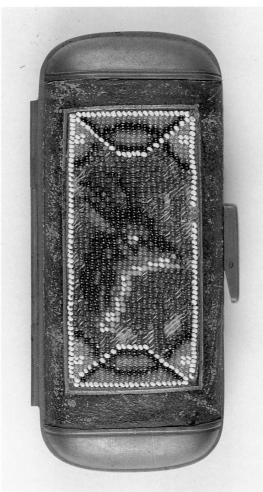

Tooled and tinted leather cigar case. 6" x 4" closed. $275-300.

Metal and leather cigar case with bead work decor. Hinged, 5.75" x 2.75". $250-300.

Above and right: Hinged leather cigarette case with tortoise shell and brass interior. 3.5" x 4.25". $225-250.

Leather and metal cigar case with needlework inset and an unusual brass cannon latch. 5.5" x 2.75". $275-325.

Leather sliding cigar cases. 5.25"-5.5". $50-75.

Leather sliding cigar case with sterling trim and monogram. 5". $100-150.

Hinged leather cigar case with metal trim and a cutter inside. 5" x 3.75". $125-150.

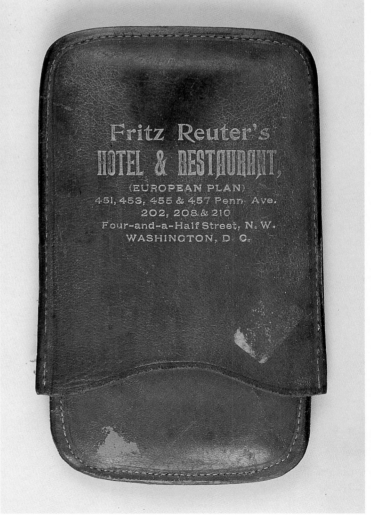

Leather slide cigar case with advertising for a hotel and restaurant. 5.25" x 3". $45-60.

Pressed paper cigar case advertising Colonel North cigars. 5.5" x 4". $50-60.

Simple monogrammed cigar case in sterling. 5" x 2.25". $175-225.

Two aluminum cigar holders, one a souvenir of Buffalo, New York. 5". $40-60.

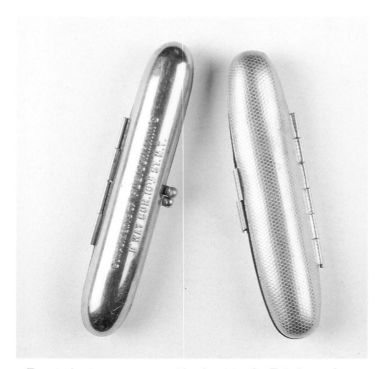

Two single cigar cases, one with advertising for Fleischmann's, New York. 4.5". $50-75.

Hinged cigar case for three cigars. 5". $75-125.

2. GOT A LIGHT?

A cigar isn't much good unless it is lighted. As we showed in our other book, there was a great variety of lighters manufactured for just this purpose. A couple of are also shown here, with their color glass globe and mechanical matches for transferring the flame.

The other popular means of lighting a cigar, of course, was with a match. To accommodate matches, beautiful matchholders were designed. Some were clev-erly designed cups that would hold the matches until they were needed. Others were mechanical, like birds pivoting down to skewer a match on their beak. Many of them show the sculptor's art and creativity.

Probably the most unusual item in this collection is not a matchholder but a coin-operated match striker. The wizard holds a match like a magic wand until a coin is dropped into the slot. Then he turns around, lighting the match against the podium-like striker.

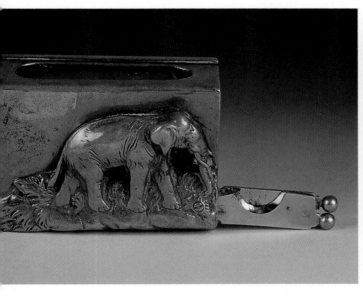

Brass matchbox cover with cigar cutter. 2.25" x 1.75". $250-275.

Brass matchbox cover with cigar cutter. 2.25" x 1.75". $250-275.

Match box cover. 0.75" x 2.25" x 1.5". $175-200.

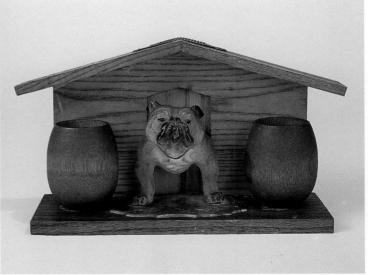

Bull dog matchholder. The house is of wood and the dog is painted cast metal. A striker is on the peak of the house. 4.5" x 7.25" x 2.75". $75-100.

Gilded bronze matchholder on a marble base. The sign is the striker. 5.5" x 5.75" x 3.75". $400-450.

Cast bronze matchholder and ashtray combination, made by L. Straus & Sons, New York. 4.75" x 5.5" d. $150-175.

A pair of cast bronze riding boots make up this handsome matchholder. 3.25" x 3" x 3". $150-175.

Cast matchholder in bronze with brass cup and brass wire laces. 2.25" x 4.25" x 1.75". $150-175.

Bull dog head match holder in cast bronze. The hat lifts to reveal the matches and there is a striker on the back of the head. 3.25" x 3" x 3.5". $275-300.

Small match holder in the form of a laundry basket and washboard striker. Cast metal, 1" x 1.25" x 2.75". $175-200.

A bear bearing matches in cast bronze. 5.5" x 6.5" x 4.25". $450-500.

The frog and the stump, cast metal matchholder. 2" x 6.5" x 4". $300-350.

Tooled brass fireplace match box holder, with room for two boxes at the top. 4" x 3.75" x 2". $225-250.

Cast iron match box holder and ashtray with a nicely executed Cheshire cat in polychrome. 5" x 4" x 3". $275-300.

Cast bronze matchholder with woodpecker. 3.5" x 4.75" x 5.75". $350-375.

Cast matchholder with woodpecker. 3" x 4.5" x 2.75". $350-375.

Metal cat in a boot matchsafe. It has a porcelain liner so it also can be used as an inkwell, but the striker on the bottom reveals its true nature. 3" x 4" x 2". $250-275.

Inlaid wood and brass coin-op match striker. When the wizard is wound-up and a penny placed in the slot, he turns and lights a match against the striker. Marked Merkel, Paris. 8" x 6" x 4.25". $2500-2750.

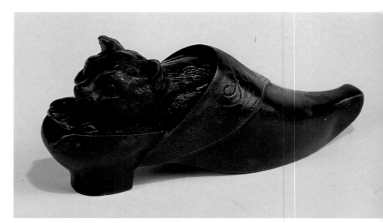

The cat comes off the slipper to reveal the matches. The underside of the cat is the striker. Marked 1920. Cast iron. 2.25" x 5" x 1.5". $250-300.

Bronze organ grinder cat matchsafe. Finely executed, the lid of the organ is hinged to hold the matches. The underside of the lid has slots to hold the striker from the match box. 5" x 3.75" d. $475-550.

Right: Yellow breasted bird guards a nest of matches. Polychrome bronze. 3" x 3.25". $225-250.

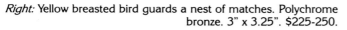

The cat sits atop this bronze mousetrap matchsafe. Behind the bars, though difficult to see are to nicely sculpted mice. The top of the house is hinged to expose the matches. 3.75" x 3.25" x 5". $700-900.

The boy is breaking out of the fireplace in this cast bronze matchholder. Signed LeBlanc, the hinged top lifts to reveal the matches. 3.5" x 4" x 3" including the marble base. $475-500.

Bisque nodder matchsafe. The top of the box lifts off. 7" x 3.5" x 3". $550-600.

Below: Ceramic matchsafe with nice cigar graphics on the lid. Two-piece, 1.75" x 4.25" x 2.5". $175-225.

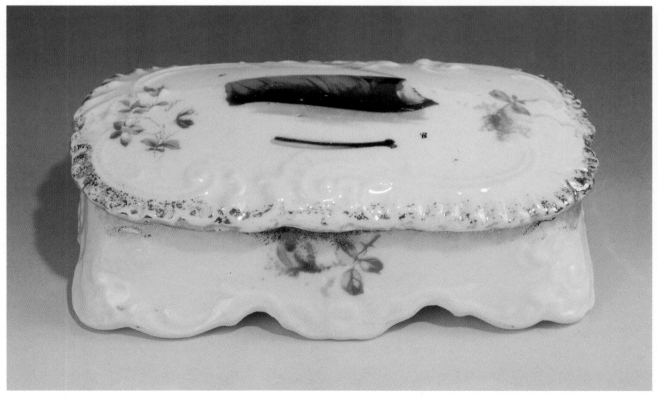

Cast iron pumpkin match holder with striker. 3.75" x 4" x 4". $175-200.

Wonderful cast matchholder in the form of the Cosmos Radiant Fire fireplace insert. The stove was manufactured by a company with the initials MV which are on the hearth plate. That plate lifts to reveal the striker. 4" x 4.5" x 2.5". $250-275.

The hinged head of this fancy cat matchsafe opens to reveal the matches. Bronze, 3.75" x 2" x 2". $300-350.

Cast bronze collie matchholder. The hinged head opens up to reveal matches and striker. 4.5" x 5" d. $500-600.

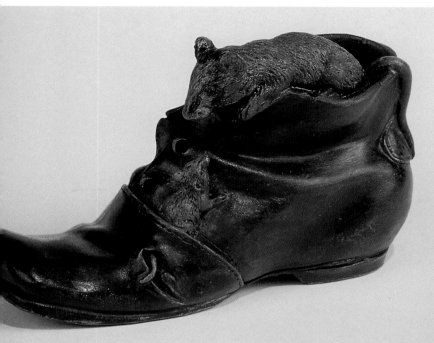

Below: Another mouse in the shoe matchholder but this is much smaller and in solid brass. Marked Peerage England. 1.5" x 3" x 1.25". $175-200.

Ceramic matchholder with nicely sculpted mice running through the shoe. 3" x 5" x 2". $175-225.

Below: Bubble crystal matchholder with etched striker. 3.5" x 3.5". $175-225.

Silverplated scissors cigar cutter. 1.75" x 7.5" x 2.5". $200-250.

Alabaster smoking set including an ashtray and a cigarette container/lighter combination. Ashtray 4.75" diameter, cigarette container/lighter 6" x 2.75" d. $75-100.

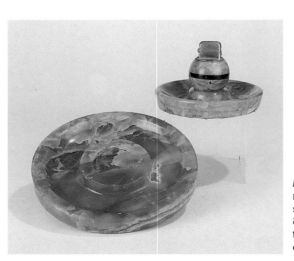

Marble ashtrays and lighter set. The lighter is in a globe of green marble with an inlaid black band. It sits in the center of the smaller ashtray. Under the smaller ashtray is the label of Halpert & Fryxell Opticians, New York, who are probably the retailers of the piece. Large ashtray: 8" diameter; small ashtray: 5.5" diameter; lighter: 3" x 2.25" diameter. $75-100.

Electric cigar lighter activated by pushing button in neck. 5.5" x 6" x 3.5". $1000-1200.

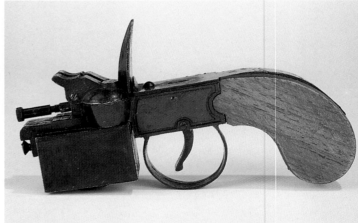

Novelty lighter, metal and wood, circa 1950. It lights by pulling trigger. 3" x 6". $50-75.

Cast metal kitten cigar lighter. The kerosene lamp at the left has a wick at the top. Marble base. 3.75" x 4" x 2". $175-200.

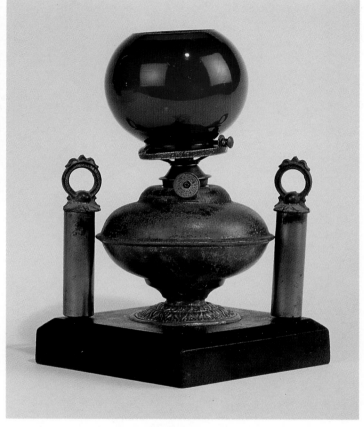

Metal and brass lighter with dip sticks, on wood base. 8" x 7". $350-400.

Beautiful table lighter in deeply embossed sterling. Marked S&F and KW. 3" x 3.25" x 1.25". $???

Nice Ronson Crown silverplate table lighter. 2.5" x 3" x 2". $????

3. ASHTRAYS

The topic of ashtrays could fill a book of its own. We have chosen a few examples that are particularly related to cigars. They range from ceramic ashtrays, with graphics that celebrate the cigar, to those that are figural and richly embossed. One goes so far as to advise the smoker that it is for "ashes." Also included are four smoking stands in a variety of designs. During much of this century a house would not have been considered completely furnished without at least one of these.

Porcelain ashtray with a pretty good poker hand. Marked Bavaria. 1" x 5" x 5". $150-175.

Ceramic ashtray and matchholder combination. Marked Austria. 3.25" x 4.75" x 4.25". $100-125.

Ceramic ashtray with good graphics including a box of cigars and a cigar cutter. Marked Victoria, Austria. 0.5" x 6.25" x 4.25". $150-175.

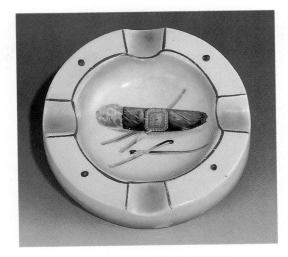

Ceramic ashtray. 1" x 4.5" d. $100-125.

Opposite page: Kerosene lighter on barrel shaped fuel reservoir, with two dip sticks. 10" x 6.5" x 4". $1500-1600.

Metal ashtray with bronze polychrome nodder on the top. Marked Austria. 4.5" x 3" x 4". $150-175.

Cigar ashtray in ceramics. Marked James M. Shaw & Co., New York, 1924. 0.75" x 4.25" x 6.5". $100-125.

Below: Tramp art metal cigar holder and ashtray. 3.25" x 6.75" x 3". $100-150.

Beautiful cast bronze ashtray. The mark of JF & Co. is on the back. It sits on four feet. 1" x 4.5" x 5". $100-150.

Copper and metal ashtray with cutter on the top. The bowl of the ashtray opens to drop the ash below. 3" x 4.5" d. $35-40.

Plated ashtray with deep embossing. The smoke from the cigar spells ashes. 3.75" diameter. $75-125.

Bradley & Hubbard smoking stand with ashtray and match-box holder. 29" x 12". $175-225.

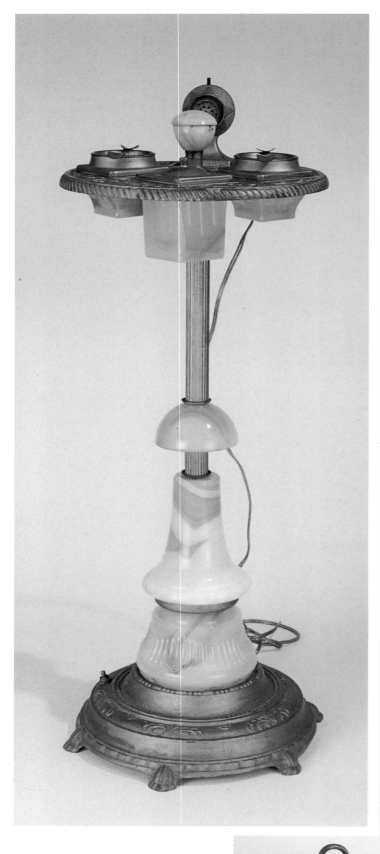

Far left and left: Metal and glass smoking stand with electric lighter marked MICO. The knob is a replacement. 29" x 11" d. $300-350.

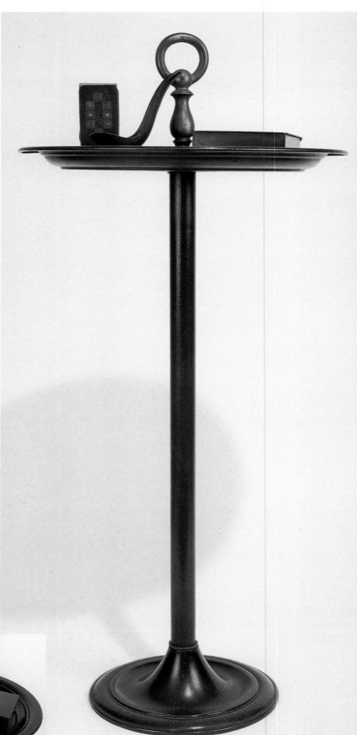

Far right and right: Smoking set from Bradley & Hubbard. It includes the ashtray/stand, a pipe rest, matchbox holder, and cigarette box. 27.5" x 13". $475-500.

56

4. MOUTHPIECES

One's style of smoking a cigar is a strictly personal thing; there is no correct way. Some prefer to chomp on the end of the cigar until it is raw. Others are more genteel, but still like to have the natural leaf against their lips. Still others like to smoke using a mouthpiece or holder. We have them to thank for some of the beautiful creations found here. Carved meerschaum, rich amber, inlay work, and filigree examples show the creative possibilities of the mouthpieces. They please the eye while enhancing the smoking experience.

Ashtray stand with lighted base. Cast metal, 28" tall. $450-500.

Cigar holders were offered in retail outlets on these cards. The card measures 11" x 6.5" and contained 10 holders, either assorted or of one style. Card with holders, $125-150.

Mint card of burl cigar holders. $125-150.

The variety of holders offered by the company. The black holders are a hard resin like Bakelite, as are the tips of the wooden holders. Wooden holders: 2.75"; black holders: 2.25"-2.5". $10-15

Whiz Bakelite Cigar Holders on their original card. Whiz was a division of Eastern Briar Pipe Co., Brooklyn. 10.5" x 9". $150-175.

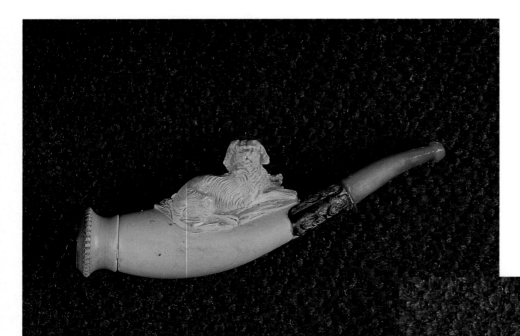

Meerschaum cigar holder with nice dog carving. 4.5" long with Bakelite mouthpiece. $150-175.

Carved meerschaum cigar holder.
3.25". $100-125.

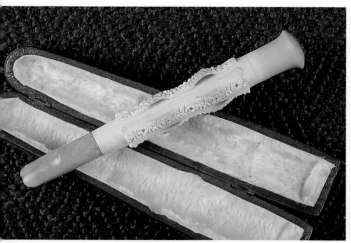

Carved meerschaum cheroot or cigarette holder with lined leatherette case. 5.5". $150-175.

Transparent Bakelite cigar holder in lined and labeled case. 4". $50-75.

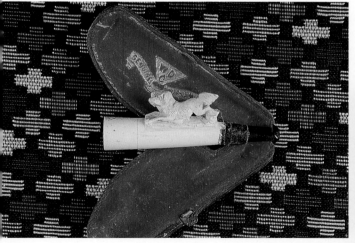

Meerschaum cigar holder in lined case. W.D. Co. 3.25". $150-175.

Bakelite cigar holder with case. 4". $50-75.

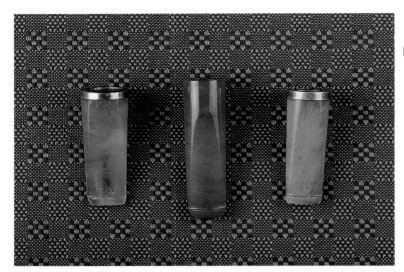

Bakelite cigar holders. 2" - 2.25". $35-50 each.

Inlaid amber cigar holder, by Windsor. 2.5". $125-150.

Meerschaum and amber cigar holders. 2" -3.25". $65-75.

Sterling clad Bakelite cigar holder (right) and enamel inlaid holder (left). 2" - 2.5". $125-150 each.

Carved meerschaum cigar holders. 4". $125-150 each.

Large meerschaum cigar holder. 4". $75-90.

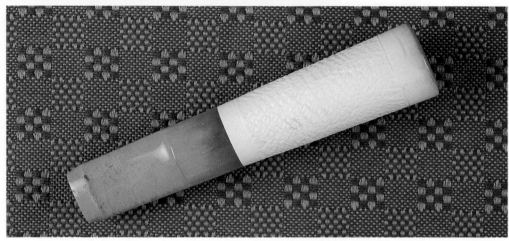

Bakelite cigar holders with metal trim. 3". $50-75.

Cigar pipe made of briar in Italy. Filled with pipe tobacco, it smokes like a cigar. 5.5". $35-50.

5. CIGAR BAND ART AND OTHER MISCELLANY

When Andy Warhol painted a Campbell's Soup can he took an ordinary object and transformed it into a work of art by forcing us to see it in another, new way. It was nothing new, however; in fact it is one of the characteristics of folk art to use ordinary objects in new and creative ways. So it is that cigar band art was born. Cigar bands, are of course miniature works of art in themselves. Beautifully lithographed and gilded, many reveal exquisite skill in portraiture and illustration. They are, included in many collections for their own sake. But some folk artists went one step further, combining many of these bands to make colorful designs. Glued to the back side of clear glass bowls, plates, glasses, or trays, they make up intricate, colorful patterns, often following a theme, such as the presidential theme shown below.

Cigar band deep saucer or ashtray. 1" deep x 5" diameter. $50-65.

Cigar band tramp art was a very popular use for these little graphic gems. Most of the art has felt backing, as does the inside of this glass. 3.75" tall. $35-45.

Cigar band deep saucer or ashtray. 1" deep x 5.5" diameter. $50-65.

State names and presidents seem to be underlying themes in this cigar band ashtray. 1.25" deep x 6" diameter. $75-85.

A larger cigar band plate. 1" deep x 7.5" diameter. $50-65.

Uncle Sam and most of the Presidents from George Washington to Theodore Roosevelt decorate this large cigar label bowl. 9" x 2". $125-150.

Nieman-Marcus picked up the cigar theme with these tumblers with their Montecristo Habana cigar labels. Marked Nieman-Marcus, circa 1955. Each tumbler is 4" x 3.25" d. $75-100 set.

Below: Lithographed fabric tribute to the charms of tobacco. 22" x 22". The quotes are of interest. $250-275.

Top left: "For Maggie has written a letter
to give me the choice between
The wee little whimpering love.
and the great god Nick O'Teen"

Top center: "Sweet black briarwood pipe of mine
If you were human you'd be divine.

Top right: Take your worms and fishing pole,
and a jug along for health.
An' you'll get a taste of heaven
From the pipe you made yourself.

Center left: "There's a lot of solid comfort

in an old clay pipe I find.
If you're kind of out of humor
Or in trouble in your mind.

Center right: "And so farewell, a long farewell
Until the weddings o'er
And then I'll go on smoking thee
Just as I did before.

Bottom left: Keep me at hand and as my fumes arise
You'll find ajar the gates of paradise.

Bottom center: She's my idol, she's my queen,
is my lady Nicotine.

Bottom right: For a woman is only a woman
But a good cigar is a smoke.

Cast iron cigar store Indian figurine or paper weight. 6.75" tall. $275-350.

Opposite page: A beautiful cigar band tray with a calendar girl in the middle. She bears the date 1917. Wood tray with glass inset. Overall length 19.75", width 11". $125-150.

Above and right: While technically a cigarette dispenser, it is so nicely done we had to share it with you. The box is beautifully inlaid wood. Cigarettes are lifted to the top when the lid is opened. 3.75" x 6" x 4.25". $175-200.

Cigar fans. Two types, both 5.5" long closed. $25-35 each.

Recruit Little Cigars flag cards. One came with each box of 10 cigars. The author received them stored in a Recruit box. $75-100.

6. HUMIDORS, CABINETS, AND DISPLAY CASES

While the merchant is certainly concerned about keeping the cigar fresh, his first job is to make it appealing to the customer. Here we have a wide range of display items for cigars. They range from beautifully constructed cabinets and massive lithographed tin humidors, to tin and glass display humidors and glass cigar box covers that slipped over the cigar box to allow the customer to see the product while keeping the air out.

Left and above: Beautiful cigar storage cabinet for Morten & Co., New York, importers of H. Upmann Habana cigars, circa 1880. 17" x 12.25" x 9". $1500-1650.

Left and below: Mahogany cabinet to hold boxes of H. Upmann cigars, circa 1890s. 21.25" 22.5" x 9.5". $600-700.

Above and left:
Large Buck Cigar
humidor. 18" x 27" x
20". $900-1000.

Large Cremo store humidor. Lithographed tin. $900-1000.

Left and above: Brass humidor for L.A. Palina Senators, with raised letters on the knob. Marked "Made of Chase Brass" on the bottom. $175-200.

Below: Beautiful brass and wood humidor with metal lining, manufactured by Grammes of Allentown, Pennsylvania. 3.5" x 10.5" x 7.75". $300-400.

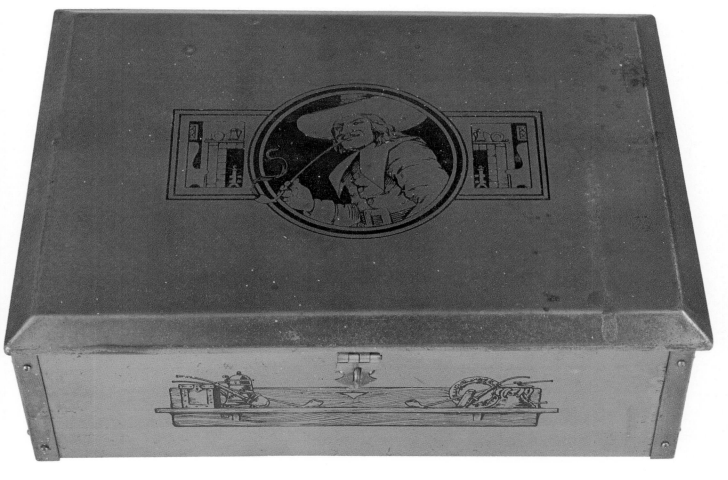

Above: H. Upmann glass humidor jars with metal fittings. These two jars are virtually identical except that one retains its cedar lining. Both have Cuban certificates. 8" x 6" d. $50-75 each.

Left: Pressed glass cigar canister for Mercantile Cigars. The company name is molded into the knob of the top and Factory No. 504, 1st Dist. MD. is molded into one of the side panels. $225-250.

Metal and glass display case for F.L. Hommer Fine Cigars, St. Joseph, Missouri. $1000-1250.

Counter display box for Historian Cigars. Glass, metal, and cardboard. 5.5" x 9.5" x 6.75". $100-125.

Wood and glass display case for Valenta Jamaica cigars. The label on the back identifies the cigar manufacturer as Martins, Ltd, 62 Piccadilly, W.I. 12" x 16" x 20". $350-400.

Glass and metal display box for Pippins and Harvard cigars. Bags for the cigars, seen at the right, are held at the back of the box. When the lid is lifted a bell sounds, letting the clerk know that someone is buying. 9" x 7" x 14". $350-400.

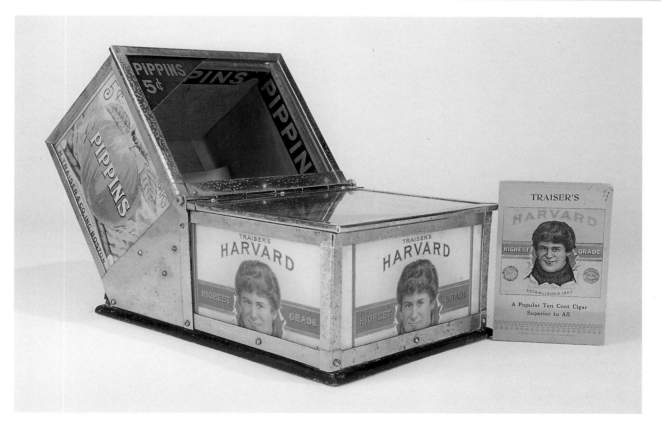

Double box tin and glass display cases for the Arabela cigar. Made by Tudhope Metal Specialties, Orillia, Canada. 13" x 10.5" x 9". $375-400.

Above: Tin and glass display case for Robert Burns Cigars with tin showcard at top. 16" x 10" x 10.5". $375-400.

Below: Two-tiered cigar display box of tin and glass, advertising Punch Cigars. Made by Tudhope Metal Specialties, Orillia, Canada. 13.25" x 10" x 13.75". $375-400.

Double box display case, tin and glass made by Tudhope Metal Specialties, Orillia, Canada. It advertises Bachelor 10 cent and Stonewall Jackson 5 cent cigars. 13" x 10.5" x 9". $375-400.

Two box display humidor for Bold Cigars by Bobrow Brothers. Tin and glass with a wood bottom. The humidifying element goes behind the sign. Made by Brunhoff Manufacturing, Cincinnati. 14.5" x 10" x 15.25". $250-300.

Countertop display humidor advertising Garcia Grande Cigars. Made by Brunhoff. 15.25" x 10.25" x 19". $250-300.

Councilman Cigar display humidor. Tin and glass with wood base, 16" x 10.75" x 21". Made by Brunhoff. $250-300.

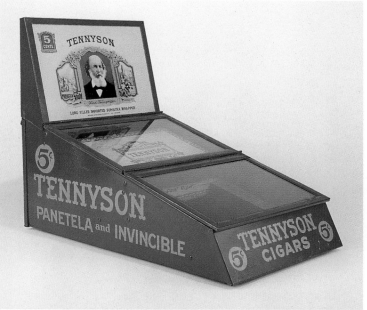

Tin and glass display humidor for Tennyson Cigars. Manufactured by Cadillac Can Company, Cincinnati. 12.25" x 9.25" x 15.75". $250-300.

Above and right: Glass and lithographed tin display humidor for Manuel Cigars. At the back is a bracket that holds the cigar box top upright for more advertising. The humidor was manufactured by the Ludlow Mfg. Co., Ludlow, Kentucky. 4.75" x 10.25" x 7.5". $75-100.

Cigar box display lids for Henry Grauley, Cinco, Tennyson, and Canadian Club. Tin and glass, 1" x 8.25" x 6". $50-75 each.

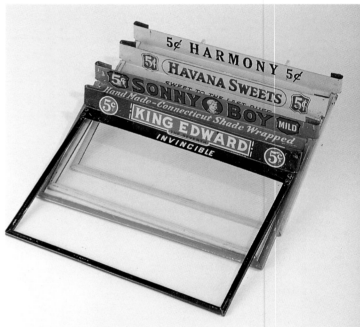

Cigar box display lids for King Edward, Sonny Boy, Havana Sweets and Harmony cigars. $50-75 each.

Cigar box display lids for After Dinner, Cremo, 44, and White Owl cigars. $50-75 each.

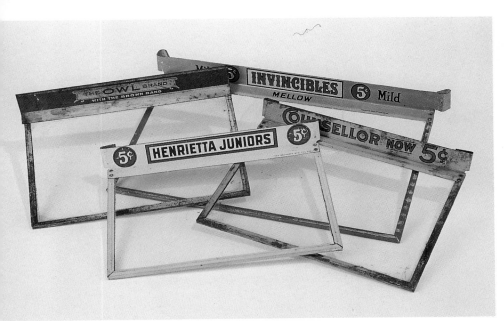

Glass and tin display lids for Henrietta Juniors, Counsellor, Owl, and Invincible cigars. $50-75 each.

Glass and tin display lids for Orange Blossom cigars. These three show a variety of styles. On the left is etched glass, top right is reverse painted glass, and bottom right is lithographed tin and plain glass. $50-75 each.

The 7204 and King Oscar display lids are etched reversed painted glass. The Victor lid is lithographed tin. $50-75 each.

The glass display box lid as it would be used.

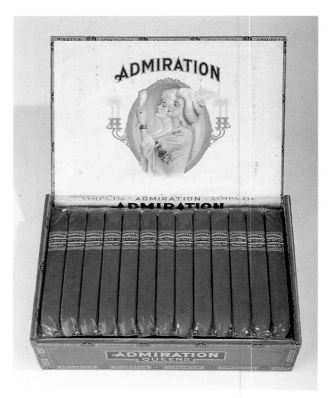

Above and below: Display box for Admiration Cigars. Inside is a "cigar plaque," cigar wrappers made to look like a full box of cigars. The unit was made by Adartcraft, Inc., New York. $125-150.

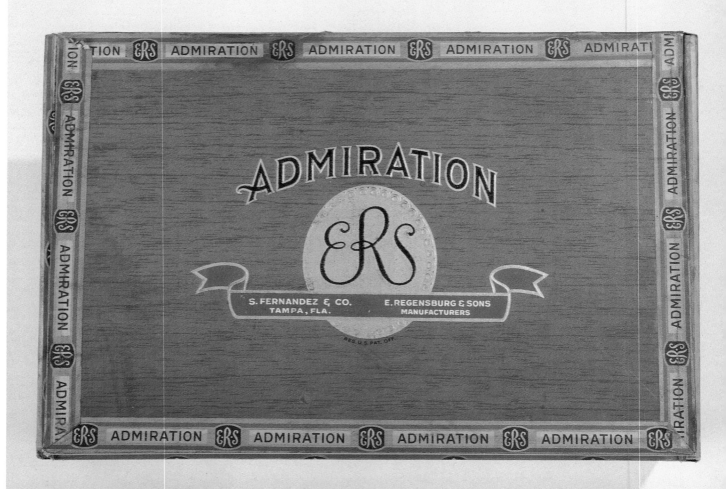

Victory cigar display lid for a tin of 50 cigars. $125-150 (lid).

Display lid for tin of 50 Orange Flower cigars. Tin and glass, 10" x 8". $125-150 (lid).

This lithographed tin wrap went around the base of a box of cigars to make up a display unit. Advertising Havana Fives, it was made by Liberty Can Co., Lancaster, Pennsylvania. 2.75" x 4" x 4". $50-60.

A box of cigar envelopes, for Reliance Cigares, Paris, circa 1920-30. The paper sleeves opened up to hold the customers cigar after purchase. The box contains 100 sleeves and is cardboard, 1" x 5" x 4.5". $60-75.

7. COIN-OPS

The history of the coin-operated machine dates back to the mid-nineteenth century. From games of chance to mechanical dispensers of products they were so popular that no self-respecting general store would be without one. The cigar dispensers came in many forms, each of them showing the inventive ingenuity of its creator.

The Elm City coin-operated cigar dispenser, shown below, was licensed in 1893. The machine was made by John Bradley of the New Haven Car Register Co. By inserting a coin at the top one could retrieve a cigar at the bottom.

Below it is a game of chance, which gave cigars as a prize for winning hands. To win one cigar you needed a pair of Jacks or higher, or three of a kind.

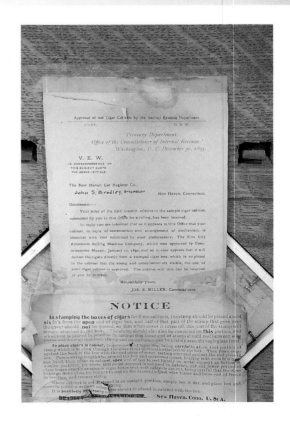

Above, top right and right: The Elm City coin-op cigar dispenser, licensed in 1893. The machine was made by the New Haven Car Register Co, John S. Bradley, successor. The label of John S. Bradley, New Haven appears on the front of the machine. 29.5" x 14.5" x 11". $3500-3750.

Above: Tin coin-op match dispenser, advertising Lucky Strike Cigarettes. 13.5" x 10.5" x 7.5". $350-400.

Left: Oak and metal coin-op trade stimulator. The prizes were given in cigars based on the poker hand that came up in the window. Of unknown manufacture, this is a truly beautiful piece. 14.5" (overall) x 10.5" x 9.75". $1750-1850.

Coin-operated cigar dispenser. Tin and glass. 9.5" x 6" x 16". $1100-1250.

Left: Generic countertop cigar machine. Cigaromat Corporation of America, New York. 32" x 13" x 9". $400-450.

Phillies cigar dispenser. 29" x 8.75" x 7.75". $325-375.

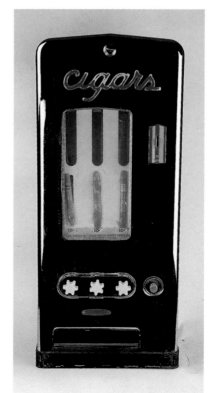

Right: All-chrome cigar coin-op machine with Bakelite knobs. Cigaromat Corporation of America, New York. 32" x 13" x 9". $450-500.

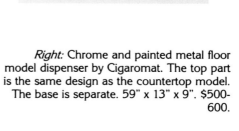

Right: Chrome and painted metal floor model dispenser by Cigaromat. The top part is the same design as the countertop model. The base is separate. 59" x 13" x 9". $500-600.

Floor model cigar coin-op called the Cigar Counter-Matic operated by Romco & Associates of Minneapolis. One piece, 54" x 8" x 8.5" (at the top). $350-400.

Generic coin-op cigar dispenser. 30" x 6" x 7". $250-275.

Left: Countertop coin-op cigar dispenser for Roi-Tan. Marked Romco & Associates, Minneapolis. 24" x 8" x 8.5". $275-325.

8. CIGAR BOXES AND THEIR CONTENTS

One wonders why a woman would allow her husband to smoke in the house. Perhaps in this earlier time a man truly did "rule the roost." Or perhaps, and this is more likely the case, she couldn't wait for the cigar box to be empty so she could use it to store sewing notions, household supplies, or any number of things. The boxes were convenient and sturdy. Made of wood and heavy cardboard, they usually had a tape-hinged lid that sealed shut with a small brad.

Besides being utilitarian, many cigar boxes had their beautiful side...usually inside. The lids were decorated with paper labels, exhibiting the finest of the lithographer's skill. While their practicality may have kept the boxes from being tossed on the rubbish heap, it is the beauty of the graphics that is cherished today.

Occasionally these old boxes still have old cigars in them. If the cigars have been cared for in the proper way and have not become moldy from too much moisture or flaky from too little, these old smokes are still smokable. Indeed some of them have aged like a fine wine, with both their taste and their value growing over the years.

Wood box with finger joints for Dubonnet Habana De Luxe Coronas. It contained 10 cigars. Factory No. 132, 3rd District New York. 1.5" x 6" x 3.5". $25-35.

Left and above: Wooden cigar box marked "Happy New Year! 1887," from El Danubio Cigars. The label inside the lid has a white cat. Factory No. 1580, 3rd District New York. 25 cigars. $50-75.

84

Above and left: Wooden salesman's sample box, marked Back's Nutura Cedarap cigars. Inside are the variety of sizes offered, and examples of the custom printing the customer could order on the cedar-like wrap. Marked Factory No. C 249, New York. 1.5" x 8.5" x 6.5". $400-450.

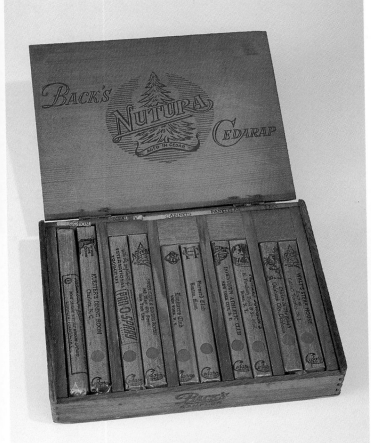

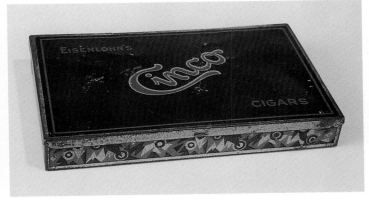

Lithographed tin Cinco Cigar box with nice deco decoration on the sides. Factory 471 District of Pennsylvania. 1" x 8" x 5". $35-45.

85

Above: Paper covered wooden cigar box for 25 Punch Havana cigars. 2.5" x 8.25" x 4". $25-35.

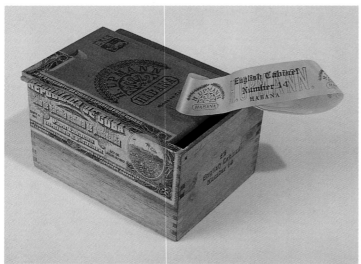

Above: Wooden slide top case for H. Upmann Habana Cigars. Dated 1952 it was used for export, having the Republica of Cuba seal on the case. 3.25" x 5.5" x 4". $35-40.

Right: Benson & Hedges cigar box for 25, circa 1933. Wood and paper. 1.5" x 9.25' x 5.5". $25-30.

Above and right: Wood, cardboard, and paper cigar box for Cremo. 1.25" x 9" x 5.75". $25-35.

Above and left: Wood, cardboard, and paper La Palina cigar box, Congress Cigar Company. Contained 50 cigars. 2.5" x 8.75" x 5.5". $25-30.

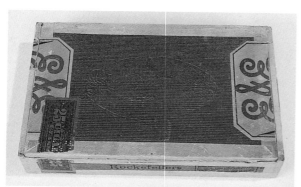

Above and right: Wood and paper box for Estabrook & Eaton Rockefellers. The outside is impressed with the company name, and the label on the inside has a Southern plantation scene. 1.5" x 8.75" x 5.75". $35-40.

Above and right: Though rather plain on the outside, the label inside the Overland Cigar box has a great graphic of a locomotive. S.S. Pierce, Boston. 1.5" x 9" x 5.5". $45-50.

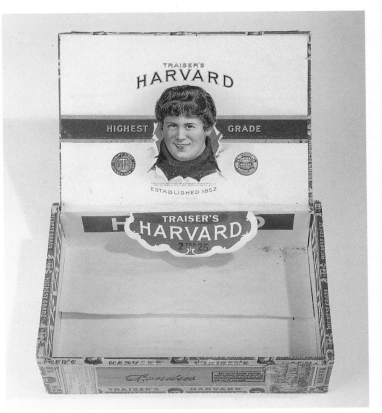

Above and right: Harvard Cigars wooden box, imprinted on the top and nicely labeled inside. 1917 tax stamp. 2.25" x 8" x 5.5". $45-50.

Below: Wood and paper Moki cigar box, with nice representation of Chief Moki. 2.5" x 7.75" x 5.25". $45-50.

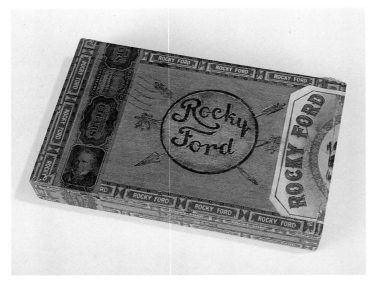

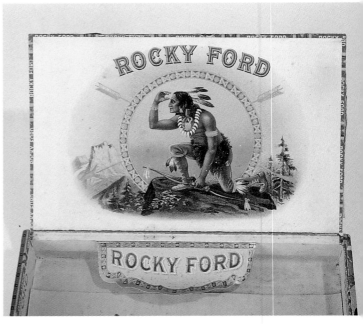

Above and right: Wood and paper Rocky Ford cigar box with painted logo on lid, and wonderful graphics. 1901 tax stamp. 1.25" x 8.5" x 5.25". $45-50.

Above and right: La Teresa wood and paper cigar box. 1917 tax stamp. 1.5" x 8.75" x 6". $25-30.

Below and right: Eden de Baces y Lopez cigar box. Cardboard, 2.5" x 9" x 5.5". $25-30.

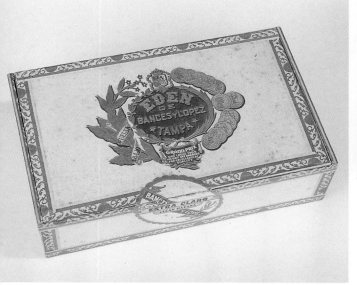

Above and left: Haddon Hall wooden cigar box. Series 117 tax stamp circa 1947. 1.5" x 8.5" x 5.75". $25-30.

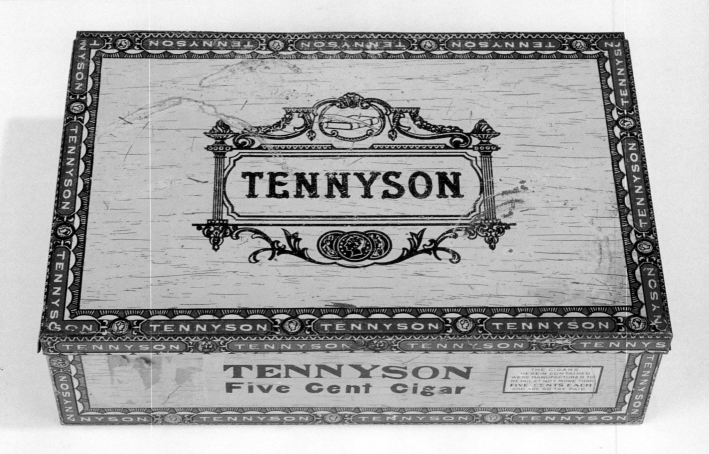

Lithographed tin cigar box for Tennysons. On the back is a nice image of the Mazer Cressman Cigar Co. building in Detroit. $75-100.

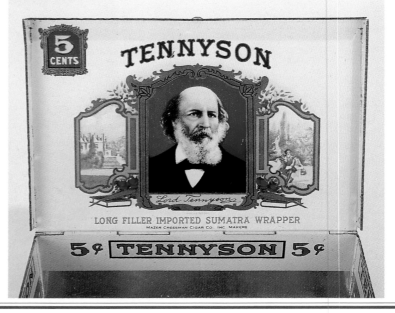

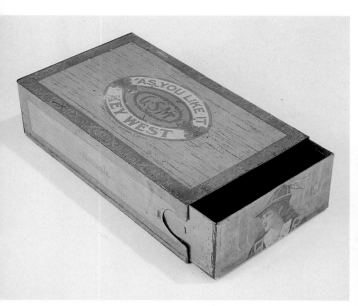

Tin cigar box with sliding drawer for Godfrey & Mann's As You Like It cigars. A 1914 patent is recorded on the lid, which is assumed to be for the box design. 2.5" x 9" x 5". $50-60.

Nicely made wood cigar box with metal hinges and clasp for Sano, A Scientific Cigar, circa 1926. This smoke was "denicotinized," having less than 1% nicotine. 2.25" x 8" x 6". $75-85.

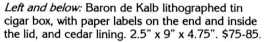

Left and below: Baron de Kalb lithographed tin cigar box, with paper labels on the end and inside the lid, and cedar lining. 2.5" x 9" x 4.75". $75-85.

Left and above: Wood Lady Churchill box with embossed logo and paper labels. 1917 tax sticker. 2.5" x 9.25" x 5.5". $45-50.

Bottom left and below: La Primadora wood and cardboard cigar box with wonderful paper labels and front flap. 1926 tax sticker. 1.5" x 8.5" x 6". $45-50.

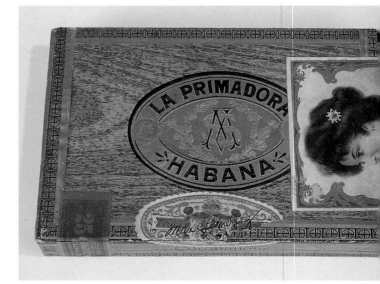

Above and left:
Cardboard cigar box for Dixie Maid, Factory Smokers, circa 1933. NRA logo on the inside of the lid. 2.25" x 17" x 5.5". $35-40.

Cardboard pocket box for 5 Blackstone Cabinets, by Waitt & Bond, circa 1926. 0.5" x 5.5" x 3.5". $25-30.

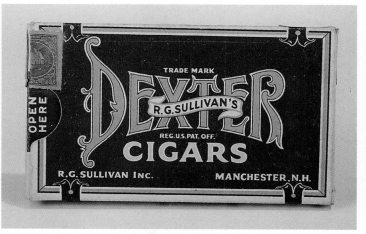

Cardboard pocket box for 5 Dexter Cigars from R.G. Sullivan's, 1952. 0.5" x 5.25" x 3". $25-30.

Above and right: Roy B. Starn Genuine Havana Stogies box and original cigars. Box, $10-20; cigars, market price.

Above and right: Cube shaped box for 100 Sixty-Six cigars. Box, $25; cigars:, market price.

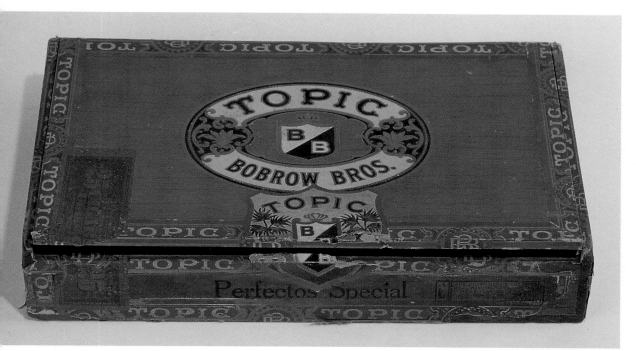

Above and left:
Topic cigars, circa 1917. Box, $25-35; cigars, market price.

Below and right:
LaRinca Cigars, circa 1946. Box, $25-35; cigars, market price.

Above and right: Mozart Superiores, circa 1951. Box, $25-35; cigars, market price.

Above and right: Cinco Cigars, circa 1926-28. Box, $25-35; cigars, market price.

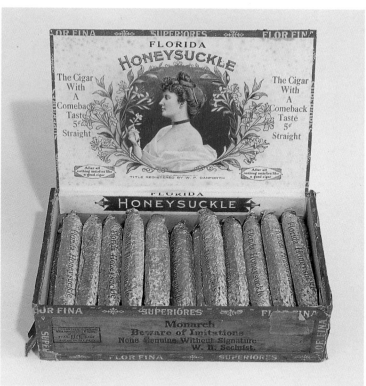

Above and right: Florida Honeysuckle Cigars, circa 1917. Box, $50-60; cigars, market price.

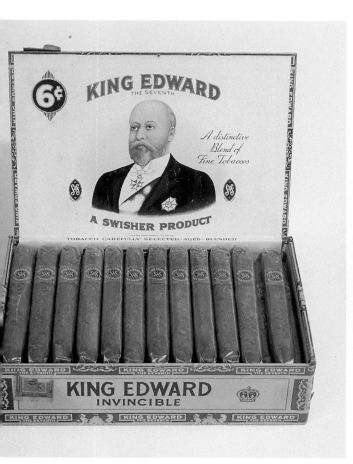

Left and above: King Edward Cigars, circa 1952. Box, $25-35; cigars, market price.

Socrates cigar in its original glass tube with the gold foil seal intact.

Orange flower cigar tin. It held 50 cigars. 5.5" x 6". $125-150.

Little Bobbie hinged tin held little cigars from the makers of Robert Burns cigars. 1" x 4.5" x 3". $25-35.

Que Placer cigar tin. 1" x 5.5" 3.5". $25-35.

Three varieties of Bayuk's Philadelphia Phillies cigar boxes in lithographed tin. Though they are not dated this is presumed to be chronological order from left to right. The tin on the left has the NRA logo dating it to 1933-1935. The others are from the 1930s or 1940s. Each tin has a prop arm that holds the lid open for display. 3" x 7.25" x 5.5". $50-65 each.

Lady Churchill cigar tin. 1.25" x 5.25" x 3.5". $25-35

Hinged Y-B cigar tin. 1.25" x 5" x 3.5". $25-35

Flat hinged tin held 10 Cinco-Nettes Little Cigars.
0.5" x 5.25" x 3.75". $25-35

El Murillo cigar tin with applied paper
labels. 1.25" x 5" x 3.5". $25-35

Above: Alles & Fisher's J.A. hinged cigar tin. 2.5" x 5.25" x 3.5". $25-50.

Below and right: Hinged Dubonnet DeLuxe cigar tin. 1.25" x 5" x 3.5". $25-50.

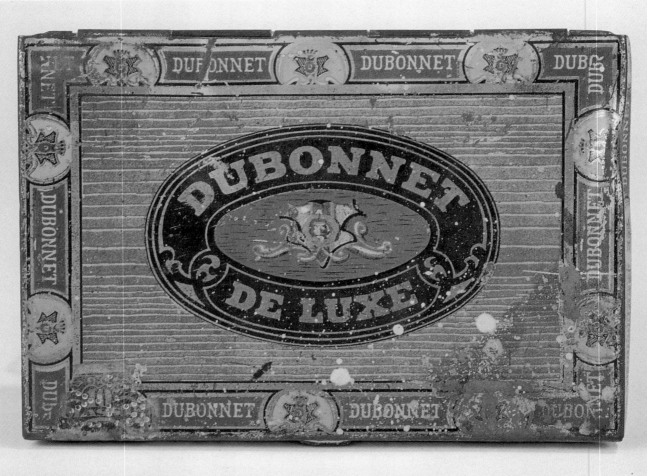

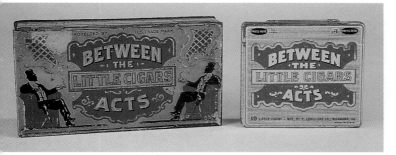

Two Between the Acts hinged tins. The larger has an embossed top and held 50 cigars, circa 1900. The smaller dates to circa 1955 and held 10 little cigars. L: $50-75; R: $25-35.

Right: Cinco vertical cigar tin for 25 cigars, circa 1916. 5.25" x 3" x 3". $50-75.

Below: Hinged tin for Prize Winners, circa 1900. The tin was manufactured by Wm. Vogel & Bros., Brooklyn. The front and back of the tin are pierced for air circulation. A moistened blotter over the holes would have provided humidity for the cigars inside. 1.5" x 4.5" x 3". $250-275.

Cigar canister for Check Cigars, Rock City Cigar Co., Quebec. The lithographed tin is by the A.R. Whittall Can Co., Montreal. 5" x 5.25" d. $200-225.

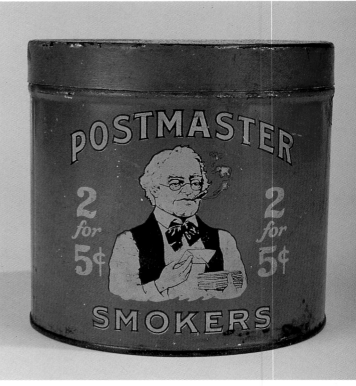

Postmaster Smokers cigar tin, circa 1900. 5" x 5.25" d. $150-175.

The Erb cigar tin. Has 1917 tax stamp. 5.25" x 6" d. $75-85.

Punch cigar tin. 5" x 5.5" d. $150-175.

Lord Tennyson Puritanos, Canadian tin with nice graphics. 5" x 5" d. $100-125.

Emilia Garcia cigar tin. 5" x 5.5" d. $50-75.

White Ash cigar tin, circa 1930s. 5.25" x 6" d. $75-100.

Above and right: Oval lithographed tin for Alvarez Lopez Y Ca Coronas, decorated all around, circa 1910. 6" x 5.5" x 4.75". $50-65.

Briefs Cigars tin. 5.5" x 3.25" x 3.25". $75-100.

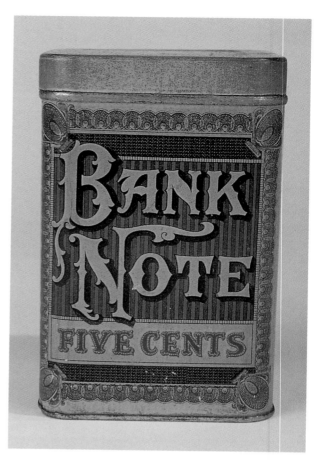

Bank Note Cigars tin. 5.5" x 3.25" x 3.25". $75-100.

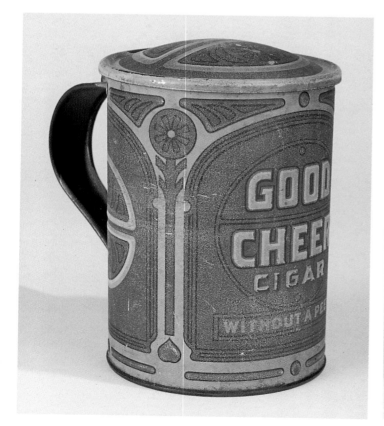

Good Cheer cigar tin in stein form. The tin is lithographed and embossed. 6" x 4.75" d. $175-200.

Beautiful cigar tin for Langaard's Portorico No. 1 cigars. While in the main the graphics depict people smoking pipes, the cigar bands on the end panels reveal the true contents of this unusual tin. Inscribed as a gift on the bottom and dated 1904. 3.5" x 6" x 3.75". $200-250.

9. ON THE COUNTER

In America the tobacco industry was at the forefront of advertising. Tobacco companies were among the first to recognize the importance of a tradename to lift one's product above the fray of the generic. They used it on packaging, as we have seen, so that even from several feet away, their particular cigar could be seen and chosen by the customer. Morever, in an age of high illiteracy, they relied on distinctive, often beautiful graphics to identify the product without words. The Harvard undergraduate, the Overland train, the Moki Indian—each had a unique visual image that could be readily chosen.

This advertising appeared not only on the packaging or signs promoting the product, but at every point where the buyer might be influenced. Lighters, cutters, box openers, statues, paperweights, change trays, and ashtrays all carried a product message. The effort to catch the customer's eye led to the creation of some remarkable objects, many of which are shown here.

Cast plaster cigar store Indian, circa 1940. 39". $600-750.

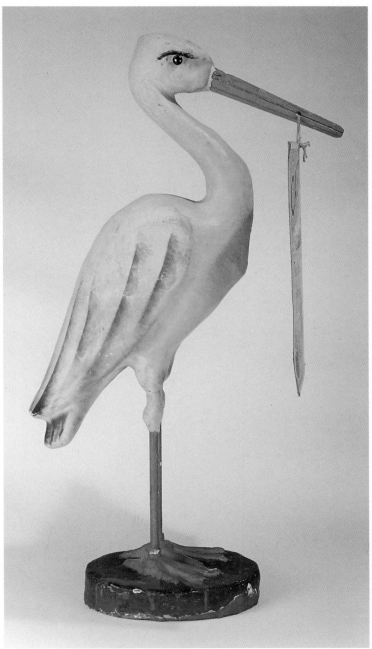

This papier mache and wood stork holds sign introducing "The New Arrivals," Castle Hall Twins Cigars. The stork is 25" tall. The cardboard sign measures 10" x 12". The sign was printed by Petre, Schmidt & Bergmann. Sign: $150-175; Stork: $500-700.

Above: Durham Cigar played off the popular logo for Bull Durham cigarettes, though the two companies are unrelated. Durham Cigars were made by Kettermann Bros. of Louisville, Kentucky. This advertising clock is of cast iron, 9.25" x 11" x 3". $1900-2200.
Left: Advertising clock for Bell Cigars. Cast iron with brass finish, made by the Regent Manufacturing Co., Chicago. The clock has a winding mechanism. 12.5" x 10" x 2". $500-750.
Below: Advertising clock for The Boston Cigar and The Peg-Top Cigar of L.O. Grothe & Company, Montreal. Cast iron, 11" x 16" x 2". $1200-1400.

Brass letter opener advertising Statesmen Cigars. 9.5". $125-150.

This embossed tin wall clip, the 19th century's version of the refrigerator magnet, carries advertising for Hoffman House Cigars. 4.5" x 2.5". $225-275.

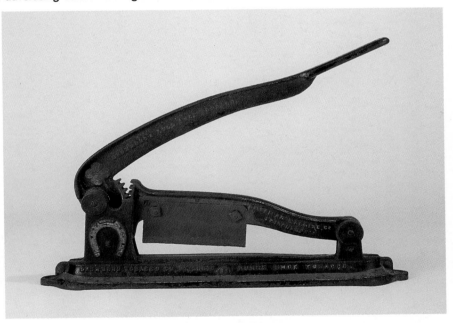

Plug cutter advertising Good Luck and Horse Shoe Tobacco. American Machine Co., Philadelphia. 16" x 3" x 6" (closed). $225-250.

Tobacco cutter advertising Cremo Cigars. 5.25" x 5.5". $100-125.

Above: Wood and metal counter lighter. This sample from the manufacturer, Knoblock-Heideman, of South Bend, Indiana, shows where the customer's advertising would go. The dip matches in the top are inserted in a slot where they are lit. 8.5" x 8.5" x 7". $700-800

Gambling device advertising John Drew Cigars. Glass and metal, 1.5" x 3" d. $125-150.

Advertising mirror for 3-20-8 Cigars by A.B. Smith. 1.75" d. $30-40.

Right: Brass printing plate for Entertainer cigars. 3.25" x 5.5". $25-35.

Wool uniform for Denby
Cigars c. 1930. $1300-1500.

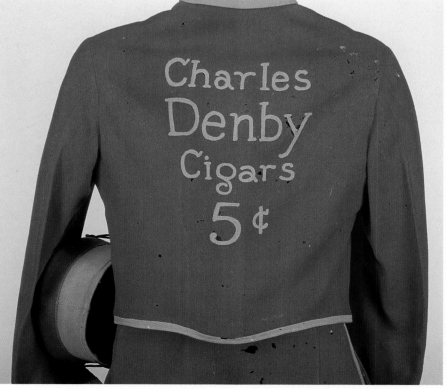

Cast metal counter display for Bull Dog Segars by John W. Merriam & Co. The Merriam Company name is in raised letters around the dog's collar. 5" x 7.5" x 5". $375-400.

Small bronze display piece for Bull Dog Segars. It is marked with Segars on its left flank and Merriam on the dog's collar. 2" x 3" x 1.5". $150-175.

Dutch Masters counter display. Cast white metal. 9.5" x 4.5" x 5". $350-400.

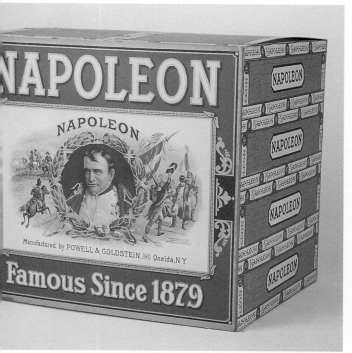

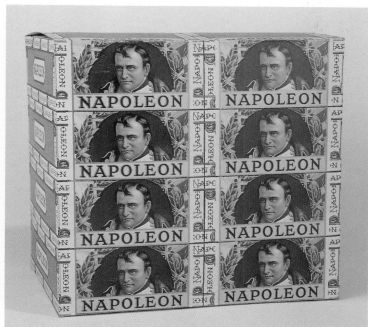

Cardboard display piece for Napoleon Cigars. 10" x 10" x 8.5". $150-175.

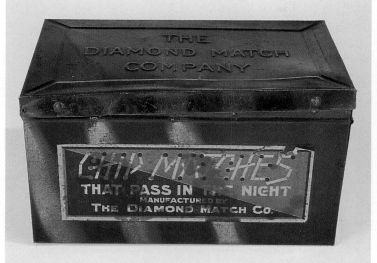

Embossed tin match box for Diamond Match Co., with paper label on the front. 5" x 8.75" x 5.5". $75-100.

Advertising match dispenser. By depressing the top, the match is lifted up for use. The decals are still intact, but the advertising on the pedestal is worn away from use. The top and pedestal are tin, the pierced base is cast iron, and it sits on a wooden base. Extended size: 5.75" x 4.5" x 2.75". $275-300.

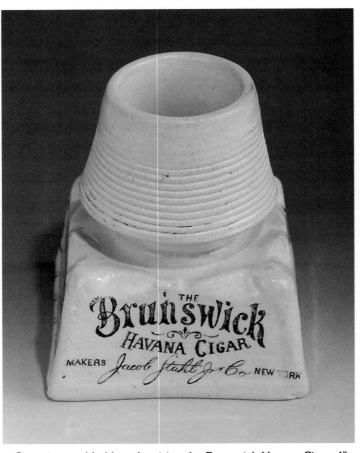

Ceramic matchholder advertising the Brunswick Havana Cigar. 4" x 3.25" x 3.25". $200-225.

Cast iron change tray advertising Napoleon and 370 Cigars by Powell & Goldstein Makers. 1.25" x 7" x 6". $125-150.

Cast iron change tray advertising Peirce's Nine Cigar. Manufactured by Brunhoff Mfg. Co., patent 799,658. 1" x 7" x 6". $125-150.

Cast iron Brunhoff change tray for Peter Pan Cigars. 1.25" x 7" x 5.75". $100-125.

116

Glass change tray with graphics beneath advertising 7-20-4 Cigars. 1.5" x 6.25" x 6.25". $75-100.

"Greely" Cigar change tray, gambling piece, used as a roulette wheel. Made by the Anderson Game Company. 1.5" x 8.5" diameter. $250-275.

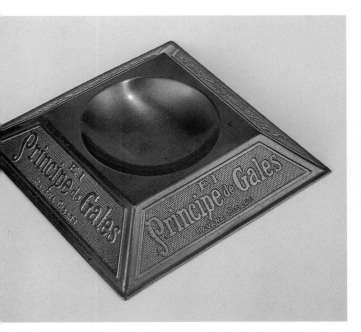

Metal change tray advertising El Principe de Gales Havana Cigars. 1.25" x 6.5" x 6.5". $75-100.

Glass change tray. The Country Gentleman Cigar graphics are pasted on the underside of the glass and painted over. Manufactured by Brunhoff. 2" x 6.5" x 6". $150-175.

Glass change tray. The Country Gentleman Cigar graphics are pasted on the underside of the glass and painted over. Manufactured by Brunhoff. 2" x 6.5" x 6". $150-175.

Girard Cigars glass change tray. 1.5" x 6" x 6". $150-175.

Fring's 3 Bros. Cigars change tray. Glass with decal and paint graphics under the glass. Made by Brunhoff. 1.75" x 6.75" diameter. $200-225.

David Crocket change tray. This is a cardboard box with metal trim and a glass insert, revealing the David Crocket graphics beneath. Patented by Benze Bros., New York. 1" x 5.5" x 5.5". $250-350.

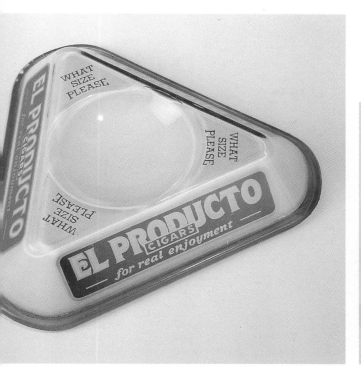

El Producto change tray, glass over paper. Brunhoff Manufacturing, Cincinnati. 2" x 9". $75-100.

Raleigh Cigarettes glass change tray on wood base, shown with its original Brunhoff box. The copy on the box declares that this is the "only practical coin tray." 2.5" x 7.5" x 7.5". $225-250.

Lithographed tin tip tray for National Cigar Stands Co., with representations of its brands around the rim. 6" diameter. $150-175.

Tip tray advertising Robert Burns Cigar. Lithographed tin. 4" diameter. $75-85.

Lord Stirling tip tray. Lithographed tin, 6" x 4.25". $75-100.

El Dallo tip tray. 6.5" x 4.5". $75-85.

Cortez tip tray, cigars "For Men of Brains." Lithographed tin, 6" x 4". $75-85.

Two tip trays for El Verso, one with the legend "Havana Cigars," the other "The Sweet & Mellow Cigar." 6.5" x 4.5". $75-85 each.

7-20-4 Cigar tip trays in various sizes: 6" x 4", 6.5" x 4.5", 7" x 5".
$50-65 each.

Above: Heavy copper and brass tip tray with embossed cigar in center advertising Lillianette Cigars. Made by S. Sternau & Co., New York. 5.25" x 7.25". $125-150.

Left: Cinco change rug. 9.5" x 12". $50-75.

Embroidered change rug for
Y-B cigars. 10.25" x 6.5".
$50-75.

Felt change rug for 7-20-4 cigars. 9.5" x 12.5". $50-75.

Footed glass change tray for Phillies Cigars. 8" x 11". $125-150.

Rubber change mat for Garcia y Vega cigars. 7.5" x 9.75". $35-45.

Rubber change mat for Muriels. 7.75" x 9.5". $35-45.

Cast iron paperweights. 3.75"-4". $125-150 each.

F.M. Kendrick & Co. Crow Cigar paperweight. The graphics are a decal applied to the underside of the glass. 1" x 4" x 2.5". $125-150.

Above: Saeger & Sons paperweight. The paper label is sealed behind the glass. 1" x 2.75" x 4'25". $150-175.

Left: White-Cat Cigar paperweight. The paper label is sealed behind the glass. 1". x 2.75" x 4.25". $150-175.

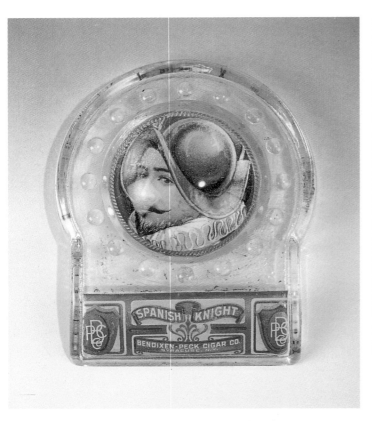

Paper under glass display piece or paperweight advertising Spanish Knight Cigars. $150-175.

Porcelain steel ashtray for 7-20-4 Cigars. 5.5". $40-50.

A pair of glass ashtrays advertising 7-20-4 cigars. One corner has a slot to hold book matches. 1.75" x 3.5" x 3.5". $25-35 apiece.

Octagonal cast iron ashtray with nicely embossed bull dog, probably an advertising piece for Bull Dog Cigars. 1" x 6" x 6". $175-225.

Cast iron ashtray advertising Bachelor or Hoffman House Cigars. 10" x 5.75". $200-250.

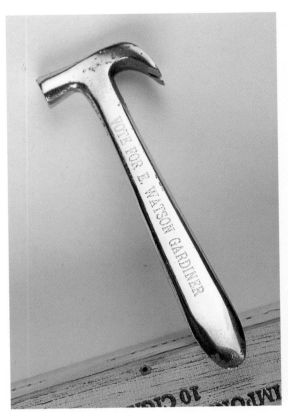

Cigar box opener used for political advertising. 4.25". $65-85.

Openers advertising cigar shops. 4.25". $65-85 each

Cigar advertising on openers. 4.25". $65-85 each.

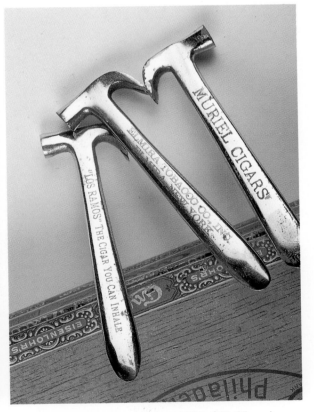

Cigar advertising on openers. 4.25". $65-85 each.

Cigar advertising on openers. 4.25". $65-75 each.

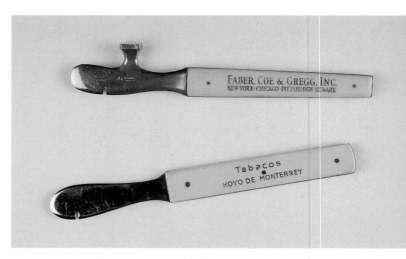

Openers with advertising on celluloid handles. 6". $65-85.

Cigar advertising on openers. 6.75". $65-75 each.

Right: Nice Polk cigar pike box opener. 6.25". $75-85.

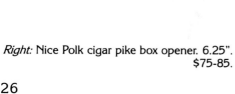

Non-advertising openers with Bakelite handles. 4.5" - 6.5". $75-100.

10. CIGAR SIGNS

The traditional way of advertising a product is to create a sign, and the walls in many country stores were plastered with them. From floor to ceiling, pillar to post they would scream out for the customer's attention. The revolution in signage, as in packaging, took place late in the nineteenth century with the perfection of color lithography. At once, what were rather plain looking, wordy black and white signs were transformed into brilliantly colored, attention grabbing examples of graphic art. They were so new and different that people began to collect them, using them to decorate the their homes and replacing the duller black and white etchings. This "color fever" is at least partially responsible for the survival of these rather fragile signs.

Not all of them were fragile, of course. Lithographed tin and porcelain enamel were not only brilliantly colored, but also had the endurance needed to withstand the elements. They come down to us in remarkably good condition, a testament to an earlier age of advertising.

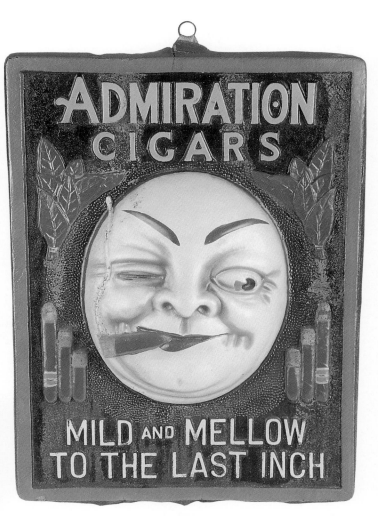

The original paper maché Admiration sign. Marked Adartcraft Inc. New York on the back. 15.75" x 12". $500-550.

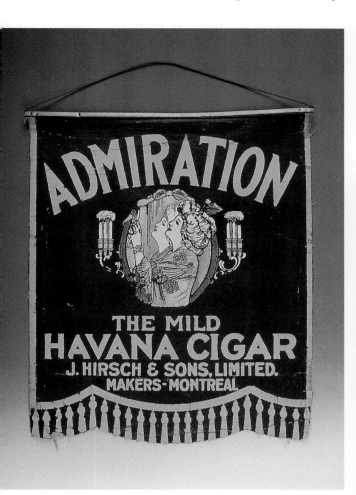

Felt hanging sign for Admiration cigars. 13" x 11". $125-135.

A plaster reproduction of the original paper maché Admiration Cigar sign. 15" x 11" x 2.5". Nice detail. $125-150.

127

A cast of the Admiration Cigar figure. It is in a concrete like material, and its origins are unknown. It may be a homemade homage to the original. The cigar belonged to the author. 9.5" diameter. $250-275.

Die cut counter sign for Airedale Cigars. Cardboard, made by Maurice Levie & Brother, Inc., Baltimore, Maryland. 12" x 9.5". $175-200.

American League Cigars hanging sign. Tin over cardboard. 6" x 13.25". $225-250.

Wonderful die cut sign from the late teens or early twenties for Bank Note Cigars. Self-standing in heavy cardboard, the bottom of the cigar box is outlined with a cut so it can be folded out for a three-dimensional effect. Banes & Mayer Litho., Philadelphia. 21.5" x 13.75". $350-400.

Lithograph tin sign for Baxter's Drum Cigar. 13.5" x 9.5". $100-125.

Cardboard counter sign for Golden's Blue Ribbon Cigars. 30" x 17". $200-250.

Mirror sign for the Beresford Cigar in oak frame. 19.5" x 24". $350-400.

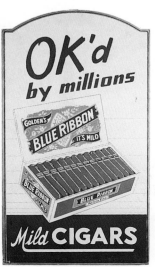

Self-standing cardboard sign for Golden's Blue Ribbon Cigars. 30" x 17". $200-250.

Golden's Blue Ribbon Cigar poster board sign. 16" x 36". $125-175.

Call Again lithographed tin sign. 3" x 13.75". $50-75.

Lithographed embossed tin sign for Call Again Cigars. Donaldson Art Sign Company, Covington, Kentucky. 11.75" x 35.25". $250-300.

Mirror sign for Buckingham, Swope & Co., Baltimore. 12" d. $300-350.

Cardboard for Canadian Club Cigars. The frame is not original. 13.25" x 21". $125-150.

Canadian Club hanger. Lithographed cardboard with graphics on both sides. 7"diameter. $25-50.

Call Again Cigar hanger. Cardboard, two-sided, 6.5" x 6.75". $50-65.

John W. Merriam produced this beautiful framed pressed paper sign in 1900. The word Segar is embossed over the Bull Dog's left foreleg. The name of the sign manufacturer is also embossed but not entirely legible. It seems to be Jones Co. Buffalo, New York. 19.5" x 16". $800-1000.

Heavy paper Capulet Cigar sign with nice graphics. 3" x 18". $35-50.

Small tin sign for Caton Cigars. 3" x 12". $50-65.

Casa-Nova Cigar counter sign. Tin over cardboard by American Art Works, Coshocton, Ohio. 8" x 11". $75-85.

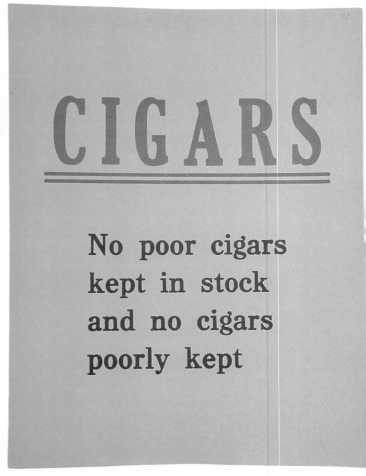

Generic store sign on heavy paper stock. 11" x 8.25". $25-50.

Castle Hall sign introducing the "New Arrival" of their twin pack. Lithography by Petre, Schmidt & Bergmann, New York. 11" x 12". $225-250.

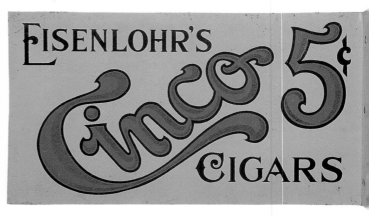

Eisenlohr's Cinco flange sign. Lithographed tin, 9" x 16". $125-150.

Cinco reverse
painted glass sign.
16.5" d. $350-400.

Colonial Club enamel flange sign. 8.75" x 18.5". $150-175.

Colonial Club Cigar sign.
Cardboard, frame added,
22" x 17". $750-800.

This beautiful, large two-sided porcelain enamel sign for Cogetama Cigares originated in Belgium, manufactured by Emailliere Belge Brux, a Brussels firm. 36.5" x 29". $2500-3500.

Wonderful cardboard hanging sign for Cyclone Twister Cigars, whose slogan is "Looks Crooked but Smokes Straight." Copyrighted 1920 by Martin L. Richards. 11" x 9". $150-175.

Window decals. Made by the Palm Brothers, Decalcomania, Colorado. 5" x 3.25". $25 ea.

Devilish Good Cigar lithographed tin sign with embossing. Sentienne & Greene, New York. 10" x 13.75". $150-175.

"Ditto" Cigars celluloid over cardboard hanger. Whitehead & Hoag, Newark, New Jersey. 12" x 8". $350-375.

DOLLY MADISON CIGAR

Block tin embossed sign for Dolly Madison Cigars. Made by National Sign Co. Dayton & Philadelphia. 5.5" x 20.25". $75-100.

Counter sign for Dundreary Cigars in celluloid over cardboard. Though unmarked, it strongly resembles those by Whitehead & Hoag. 9.75' x 7.5". $225-250.

Cardboard counter sign for El Principal Cigars. 30" x 17". $200-250

El Moriso Cigar cardboard counter sign. 10.5" x 13.5". $50-75.

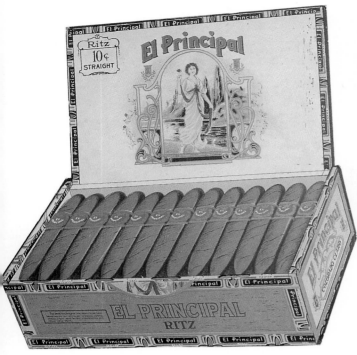

Die cut cardboard counter sign for El Principal cigars. 11" x 11". $75-100.

Paper sign for Epco-Giant Cigars. 6.5" x 20.25". $50-75.

WE SELL ERBANCO HAVANA CIGARS

Cardboard banner sign for Erbanco Cigars. Made by Miller Printing Co., Reading, Pennsylvania. 3.25" x 18.25". $75-100.

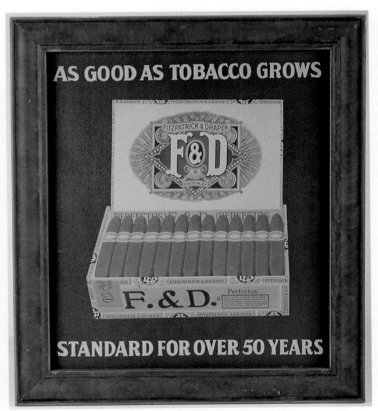

Lighted sign. The surface of the sign is translucent ink lithographed over brass screening in a process called Silicated Wire Glass, made by Meyercord Co., Chicago. $275-325.

Wonderful graphics mark this Emilia Garcia Cigar sign. Heavy cardboard manufactured by Industrial Litho Co., Brooklyn, circa 1920. 30.75" x 20". $300-350.

Framed tin sign for Francis Wilson Regalia cigars. The frame is probably original. Marked Sentienne & Green, New York. Overall size: 19.75" x 15.5". $1250-1300.

Embossed cardboard hanging 9.5" x 11.25"sign for Free Lance Cigars. The design for the sign is owned by J.G.S. & Co., Reading, Pennsylvania. $100-125.

Left: Block tin sign for Fame and Fortune and War of Wealth Cigars. 13.75" x 7.75". $125-150.

Golden Rule Cigar hanger, tin on cardboard. Ritter Mfg. Co., Philadelphia. 6.5" x 9.25". $75-100.

Beautifully lithographed and embossed original Hambone hanger. Graphics on two sides. 7" diameter. This piece has been reproduced. $75-100.

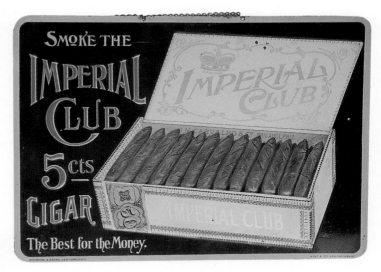

Imperial Club Cigars lithographed tin hanging sign with embossing. Sentienne & Greene, New York. 10" x 13.75". $125-150.

Hoffman House Cigars cardboard hanger. 15" x 12". $150-175.

Aluminum hanger for James Lewis Cigars. 6" x 9". $50-65.

Opposite page: Wonderful lithographed tin tray sign for Grand Duchess Cigars. The image is entitled "Chrysanthemum Girl" by W.H. McEntee, copyrighted 1910 by American Art Works, Coshocton, Ohio. 13.25" x 10.5". $450-500.

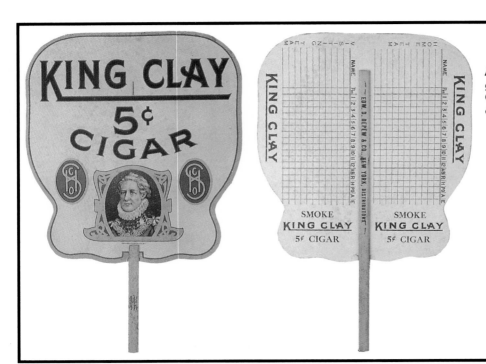

Advertising fan for King Clay Cigars. Lithographed cardboard by American Lithographic Co., New York. On the back is a baseball score card. Cardboard size: 9.5" x 8.5". $50-65.

La Creosa Cigar sign with decal on back of glass with some reverse painting. Some restoration is evident. 13.75" x 9.75". $150-175.

La Perferencia reverse painted sign. 20" x 24". $350-400.

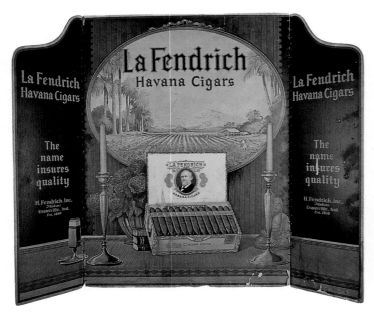

Cardboard tri-fold sign for La Fendrich Havana Segars. It bears a patent for the string hinges dated 1909. 32" x 41". $200-225.

Right: Cast iron mirror advertising Lionhead Cigars. 12.5" x 10.75". $400-450.

La Flor De Erb tin over cardboard sign. National Sign Co., Dayton, Ohio. 6.25" x 13.5". $75-125.

Right: George Raft endorses Little Fendrich's Pantellas on this self-standing cardboard sign. 35" x 25". $350-400.

144

Tin sign for M.A.C. cigars. The frame is not original. Made by the Brown Book & Seal Co., Baltimore. 10" x 14". $100-125.

Lord Baltimore Cigar cardboard sign. 6.75" x 13.25". $75-100.

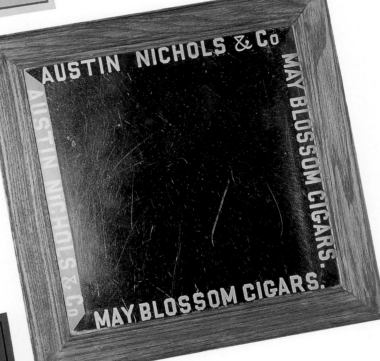

Mirror sign for Austin Nichols' May Blossom Cigars. 15" x 15" with frame. $250-300.

Lord Caspar Cigar tin on cardboard sign. Lithographed and embossed. 6.25" x 13.5". $75-100.

Napoleon Cigar lithographed tin sign. Copyright by Sanford Helmer, date unknown, probably a reproduction. 7" x 19.5".

Embossed lithographed tin sign for Napoleon Cigars. Manufactured by the Burdick Company, New York. 10" x 27.75". $200-225.

Paper sign for Old Coon Cigars. 7" x 14.5". $35-50.

Paper sign for Old Coon Cigars, made by Huntoon & Gorham, Providence. It offered a free 14k gold plated cigar cutter with the purchase of a pocket package of 12. 16" x 13.5". $125-150.

Cardboard hanger for Orange Flower Cigars. 6.75" x 11". $125-150.

Right: Paper sign for Old Coon Cigars, made by Huntoon & Gorham, Providence. It offered a free 14k gold plated cigar cutter with the purchase of a pocket package of 12. 16" x 13.5". $125-150.

Above: Peter Schuyler Cigar porcelain enamel sign. Baltimore Enamel Co., New York. 12" x 36". $200-250.

Left: Trolley sign for Peter Schuyler cigars. 11" x 21". $150-175.

QUAIL CIGARS

Above: Poster board sign for Quail Cigars. 3" x 14". $35-50.

Above: Quatility Cigars lithographed tin sign with embossed letters. 14.5" x 40". $250-275.

T.&O.Co's QUATILITY 5¢ CIGAR YES! IT IS DIFFERENT

Right: Quincy Cigar porcelain enamel sign. 12" x 36". $300-350.

BRESLIN & CAMPBELL, INC. QUINCY *Your Cigar "Q"*

RED DOT 5¢ *Truly Different* CIGAR

Left: Nicely lithographed two-sided cardboard hanger for Red Dot Cigars. 9.5" x 9.5". $50-75.

Right: Beautiful Robert Burns Cigar lithographed charger. 24" diameter. $2000-2200.

ROBERT BURNS 10¢ CIGAR

148

Embossed lithographed tin sign for John Ruskin Cigars. 9.5" x 29.75". $200-225.

7-20-4 porcelain enamel sign. 10.5" x 23". $250-300.

Porcelain enamel embossed sign for 7-20-4. 12" x 30". $250-300.

7-20-4 Cigar porcelain enamel sign. Sign made by Ingram-Richardson, Beaver Falls, Pennsylvania. 12" x 36". $250-300.

Miniature tin hanger. Two sides, lithographed tin, 5.5" x 4.25". $150-175.

Wonderful self-framing embossed lithographed tin sign for Sidney Dillon. 16.25" x 22". $1250-1300.

Block tin sign for Silver Shell Cigars. Lightly embossed in the letters. By Up-to-Date Advertising Co., Canisteo, New York. $125-150.

Decal and paint sign behind glass for Tereno Clear Cigars. Original frame. 20" x 26" with frame. $475-500.

Cardboard store sign. These generic signs left room for the retailer to add his or her own name and address. Printed by Swayze Advertising Co., Canton, Pennsylvania. 6" x 20". $50-75.

SMOKE UPTON CIGARS

MILD—FRAGRANT

Above: Upton Cigars sign. Poster board by Kemper-Thomas Co., Cincinnati. 14" x 20". $125-150.

SMOKE The WHITE LABEL 5¢ CIGARS. THE FAVORITE EVERYWHERE.

The White Label 5¢ STRAIGHT 5¢ TRADE MARK REGISTERED THE BEST GRADE AND HIGHEST PRICED 5 CENT CIGAR EVER SOLD I.N. CARVALHO & Co. PHILADELPHIA

White Label Cigars lithographed tin hanging sign with embossing. Sentienne & Greene, New York. 10" x 13.75". $125-150.

Lithographed tin license plate topper for White Swan cigars. 6" x 11". $50-75.

Right: Trade stimulator for Y-B Cigars. For 1 cent you get a chance to win a 5 cent cigar. The numbers are on little wraps of paper. Cardboard, 10.25" x 6.25" x 1.25". $200-250.

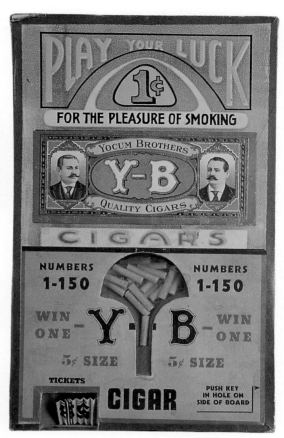

Y-B Cigar counter sign. Cardboard, 11" x 7". $100-125.

Cardboard sign for Y-B cigars. 26" x 19.5". $200-225.

11. CIGAR MANUFACTURING ARTIFACTS

Occasionally a collector comes upon some of the equipment used to manufacture cigars. These recall an earlier age, when the hand was the principle tool for making cigars. And while the best cigars are still made by hand, the industry, like most others, has turned to the robotic.

Cigar cutter used in manufacturing of cigars. Wood and metal with interesting brass spring. An identical model is seen in the 1901 Miller, DeBrul & Peters Mfg. Co., New York, where it is called the Union Tobacco Cutter. 3.75" (open) x 6.5" x 3". $85-100.

Manufacturing cigar cutter made by National Selling Co., Allentown, Pennsylvania. 2.75" (closed) x 7.25" x 3". $85-100.

Equipment used in manufacturing. At the back left is a bolted cutting board. In front of it is a cigar press, used to put pressure on molds like the one at the right. Standing on end is a cigar maker's blade and at the front right is the No. 2 Union Cigar Cutter. $500-600.

Above and right: Cigar manufacturers' record books provided by J.W Strieder Co., Boston, and Miller, DuBrul & Peters Mfg. Co., New York and Cincinnati. In addition to pages for record keeping, the books contain several pages of advertising for the cigar making products they carry. 17" x 17". $150-200 each.

155

Universal Cigar Branding & Printing Machine, J.W. Streider, Boston. The base has an automatic filling unit by Campbell & Bachellor, Boston, patent applied for 1904. Advertisements of the time claimed that the unit would brand up to 3500 cigars per hour and more in experienced hands, though they also say that it is simple enough for a child to use. 44" x 16" x 11". $1800-2000.

Brass and wood cigar printing or branding block manufactured by Gravenhorst & Co. The word is Corona. 2" x 4.75" x 1.5". $200-225.

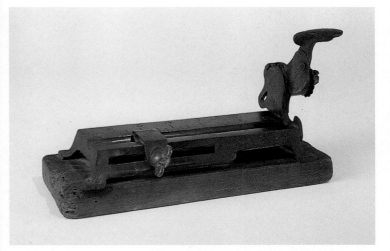

Cigar cutter, steel on a wood base. 5" x 8.5" x 3.25". $85-100.

Delegates ribbon from the Cigar Maker's Union 1896 convention. The badge is celluloid from Whitehead and Hoag. 7.5" x 3". $50-75.

1886 calendar advertising F.D. Stauffer Cigar Box Factory, Yorkana, Pennsylvania. The frame is not original. 20.5" x 18". $300-400.

12. SWEET ENDINGS

In a practice that would probably raise the hackles of the anti-smoking caucus, chocolate cigars were once a very popular candy. Made at the corner candy shop or by the big manufacturers like Hershey, they enabled children to imitate their elders, while enjoying the sweet pleasure of milk chocolate. The molds used to make these confectionery delights came in a variety of types and sizes, as you will see here. They are sought after by collectors of cigars, chocolate, and candy molds alike.

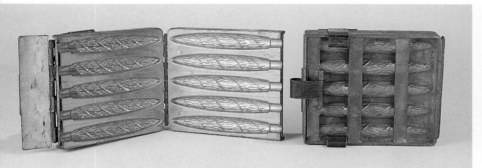

Hinged candy cigar molds shown opened and closed. Closed size: 4.5" x 5". $250-275.

Two-piece candy cigar molds, with ash. The cigars are almost 8" long. Overall mold size: 8.75" x 3.25". $250-275.

Large hinged candy mold. The cigar is 12.5" long and 2" in diameter at the tip. $300-350.

Two-piece candy cigar mold. This stogie is 9.5" long. The mold is 10" long overall. $300-325.

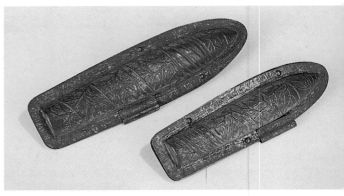

Similar candy cigar molds. The smaller is marked Tanton Reiche Dresden. Large mold: 9.5" long; small mold: 7.75". $250-275.

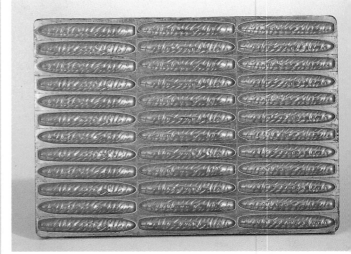

Large 36 cigar candy mold. 14.5" x 10.25". $400-475.

Hershey Chocolate produced this tray from Advertising for Hero of Manila Chocolate Segars, circa 1970. On the back is a history of chocolate. 10.5" x 13.75". Reproduction from original advertising.

Left: Two-piece mold, 14.5" long.. $300-350.